## About Island Press

# Nature Out of Place

# Nature Out of Place

## *Biological Invasions in the Global Age*

Jason Van Driesche
Roy Van Driesche

**ISLAND PRESS**

Washington, D.C. • Covelo, California

*Library of Congress Cataloging-in-Publication Data*
Van Driesche, Jason.
   Nature out of place : biological invasions in the global age / Jason
  Van Driesche and Roy Van Driesche.
     p.   cm.
  Includes bibliographical references (p.) and index.
  ISBN 1–55963–757–9 (cloth : acid-free paper)
   1. Biological invasions. I. Van Driesche, Roy. II. Title.
QH353.V36  2000
577'.18—dc21                           00–010477
                                               CIP

Printed on recycled, acid-free paper

*To my father, Roy, who has made many things possible.*
*And to Susan, who helped more than she knows.*

—JVD

# Contents

# Acknowledgments

It is astonishing how many people it takes to put together a book. We are particularly grateful to the following people, each of whom is to a significant degree responsible for the condition (and even the existence) of the finished product: Barbara Dean and Barbara Youngblood, our editors at Island Press, always had time for questions and helped us maintain an even keel. Randy Westbrooks provided helpful comments on our first outline and persuaded us that the book was worth writing. A conversation with Curtis and Kathleen Kutschenreuter early in the project gave the book its title. Ann Wieben and Willy Hutcheson each put in hundreds of hours as research assistants, and the book never would have been finished on time without their work. Gretchen Michlitsch helped clarify the argument of several of the case studies. Erhard Joeres provided institutional support at the University of Wisconsin-Madison. Stephanie Folk and Susan Munkres drew the maps. And Jennifer Ruesink, Stacy Mates, and Lloyd Loope reviewed the entire manuscript and helped us figure out how to tie the whole thing together. We are indebted to all of them.

We also owe special thanks to Dick Reardon of the U.S. Forest Service's Forest Health Technology Enterprise Team and Del Delfosse of USDA-APHIS (then of the National Biological Control Institute), who funded almost all of the several years of research, travel, and writing that went into the case studies.

Finally, we are grateful for the help of the following people who contributed data, loaned photos, participated in interviews, gave tours, reviewed chapters, or otherwise volunteered their time and expertise to the project: James Akerson, Steve Anderson, Greg Armstrong, Erik Aschehoug, Carter Atkinson, Paul Banko, Randy Bartlett, Jill Baum, Wes Beery, Pat Bily, K. Douglas Blodgett, Bernd Blossey, Barry Boerger, Steve Borror, Kathy Brock,

Bob Brumbaugh, Cathie Bruner, Gary Buckingham, Faith Thompson Campbell, Bob Carlson, Ray Carruthers, Ted Center, Avery Chumbley, Pat Conant, J. Hill Craddock, Del Delfosse, Jack DeLoach, Bill Denison, Glenda Deniston, Sam DesGeorges, Teri Devlin, Dave Egan, Lu Eldridge, Rich Evans, Tracy Feldman, Dave Fellows, Stephanie Flack, Steven Flint, Howard Frank, Cindie Fugere, Kevin Gardner, Leo Garofalo, Steve Glass, Dick Goeden, Steve Hager, Denny Hall, Dave Hanna, Chris Harmon, Henry Hart, Dave Headrick, Pam Herrera, Nelson Ho, Bob Hobdy, Mark Hoddle, Alan Holt, Woods Houghton, David Houston, Erica Howard, Frank Howarth, Kristan Hutchison, Michael Ielmini, Myron Isherwood, Jim Jacobi, Jack Jeffrey, Gary Johnston, David Jones, Bill Jordan, Lester Kaichi, Sara Kaplaniak, Brian Kitzerow, Virginia Kline, Rob Klinger, Fred Kraus, Paul Krushelnyeky, Dennis LaPointe, François Laroche, Lyndal Laughrin, Mark Leach, David Leal, Rosemary Leen, Lloyd Loope, Nancy Marks, George Martin, Nilton Matayoshi, Frank McClure, Joe Miller, Mike Montgomery, David Morihara, Patricia Morrison, Barbra Mullin, Don Mundal, Dennis Nagatani, Larry Nakahara, Richard Neves, Barbara Okamoto, Susan Pahwa, Elizabeth Painter, Chuck Parker, Henry Pierce, Noel Poe, Adam Porter, Chad Prosser, Chuck Quimby, John Randall, Dale Rapp II, Bob Ray, Don Reeser, Rusty Rhea, Ray Richard, Tim Richardson, Jenifer Ruesink, Duane Rugg, Don Sands, Don Schmitz, Peter Schuyler, Cliff Smith, Neal Spencer, Gary Sprinkle, David Stansbery, Kim Tavares, Glenn Taylor, John Taylor, Dwight Tedford, Ken Teramoto, Dave Thompson, Grady Timmons, Tim Tunison, Dirk Van Vuren, Rick Warshauer, Dick Wass, Tom Watters, Walt West, Mark Westfall, Chris Whelan, Mark White, Aulani Wilhelm, and Fred Yoder.

# *Preface*

My father and I were sitting on the balcony outside our second-floor hotel room looking out over the central square of the oldest part of Panama City. It was April of 1995. I was almost halfway through my two-year volunteer service in the Peace Corps. My family was visiting me for a couple of weeks, and we were talking about what I might do after I went home.

"I have this idea," my father said. "Do you want to write a book together?"

I looked at him.

"I'm serious," he said. "The biocontrol textbook that Bellows and I have been working on is almost done, and I've got an idea for another book— something about invasive species. Maybe for a general audience, instead of a textbook. What do you think?"

I looked at him. *Me? Write a book?* A bachelor's degree in environmental studies and a couple years of leading environmental education workshops in Panama somehow seemed a little less than adequate preparation for writing a book. Besides, I knew next to nothing about invasive species. How was I supposed to write intelligently about something I'd never really paid much attention to?

Nevertheless, we pulled out a notepad and worked out a few ideas.

*This might be interesting,* I thought.

After I came home from the Peace Corps in 1996, my father and I spent the better part of a year tossing around ideas and writing outlines. We decided to write the book in two distinct styles: first-person-narrative case studies that told stories about invasive species problems around the country, and expository essay-style chapters that provided depth and background. I was to research and write the case studies, my father would write the other half of the book, and we would edit each other's work until the whole thing spoke with a single voice. It sounded like a workable plan.

I spent most of April and May 1997 on the phone, setting up the interviews and site visits that were to form the basis of the case studies. Then, at the beginning of June, I hit the road for three months with one basic question in mind: How did each of the invaders we'd chosen affect the ecology of specific places and the lives of the people who lived there? Sixteen thousand miles and nearly a hundred interviews later, I still didn't have any answers. But I was beginning to get a sense of the big picture, and I didn't like what I saw.

We worked intermittently on the book over the next year or so. I collected articles and books, wrote outlines, and paced the house muttering to myself about latency periods and pest risk assessments. Soon my desk was overflowing with papers and notes and exclamations. The project became all consuming. All of these invasive pests added up to a catastrophe far larger than I had first imagined, and what began as a few scrawled notes on a yellow lined pad turned into the most urgently important thing I've ever done.

# Introduction

Homogeneity. Sameness. Loss of local character. This is increasingly the reality of the modern age, in which all highway exits look alike and the same stores fill the same malls everywhere. Planes and trucks and computers link the continent—and increasingly, the world—in a system that turns backwaters into thoroughfares, putting every locality in competition with the entire world. And inexorably, a few best competitors dominate one place after another until every place looks the same.

But this is the dominant trend not just in the economic world. Globalization has ecological as well as social consequences, and the same forces that are eroding the diversity of the world's cultural landscapes are to a significant degree responsible for the ongoing impoverishment of its biological diversity as well. As the growth of transportation and commerce provides more and more species with more and more opportunities to invade ecosystems far from their evolutionary homes, a relative handful of new arrivals are flourishing everywhere at the expense of a far larger number of species that are native to the places where these invaders are now found. From treekilling insects to weeds that smother entire natural communities to mussels that multiply by the billions, invasive species are among the most ecologically devastating forces that humans have ever unleashed upon the natural world.

There is nothing inherently evil about these species that have managed to migrate to and flourish in new environments. A species does not know if it is native or not. It simply does what all living creatures have always done—survive as best they can in whatever circumstances they happen to find themselves. What has changed is not the species itself, but its context. And when a nonnative species becomes invasive in this new context, the species itself is in no way to blame.

The impacts of these species-out-of-place can sometimes be quite obvious. When the chestnut blight arrived from Asia in the early 1900s and in just a few decades killed almost every American chestnut on the continent—eliminating one tree in four from New Hampshire to Georgia—everyone who read the paper or who walked in the woods knew something terrible was happening. Three generations later, anyone who keeps up with the news knows there are now billions upon billions of Asian zebra mussels blanketing riverbeds and lake bottoms across the eastern half of the United States. Invasions of this magnitude are invariably big news.

What is harder to see are the cumulative impacts of centuries of human-induced biological invasions on ecosystems all over the world. For every well-publicized catastrophe there are a hundred quiet disasters, and the stress of so many new arrivals has left many ecosystems dangerously frayed. Chestnut blight was the first of dozens of forest pests and diseases to invade the eastern United States over the course of the twentieth century, and most other tree species in these forests have since been attacked. When zebra mussels arrived in the Great Lakes in the early 1980s, they found an ecosystem that had already been profoundly altered by invasions and deliberate introductions of nonnative fish of many kinds. Though the forests are still green and the lakes are still full of water, an unending stream of invasions is changing these and many other ecosystems from productive, tightly integrated webs of native species to loose assemblages of stressed native species and aggressive invaders. In short, the earth is becoming what author David Quammen has called a "planet of weeds."

This is not the kind of world that anyone would choose to live in. It is a world where tens of thousands of species will have been decimated or driven to extinction by nonnative predators or competitors or diseases. It is a world where freshwater will be increasingly scarce because watersheds will have become tangles of weeds that use far more water than the native species they replaced. It is a world where good lumber will be expensive and rare because nonnative pests will have killed native forests and only a few species of aggressive, poor-quality invasive trees will have replaced them. In short, it is a world that will not work.

And it is a world where everywhere will look like nowhere at all. Whether the world becomes a world of weeds will be decided in the next decade or two. There is no time to waste.


This book is divided into three parts. Each part alternates between the case studies that I researched during the summer of 1997 and background essay-style chapters written by my father, Roy, that give more depth and breadth to issues raised in the case studies. The styles of the two kinds of chapters are quite distinct and are intended to complement each other. Whenever a chapter is written in the first person, it is based on what I saw and heard during my summer research trip.

Part One presents the scope and history of the invasive species problem. The story opens in Chapter 1 with a trip to Hawai'i, where introduced feral pigs are destroying the islands' native forests and the unique species they contain. Chapter 2 gives essential background on the importance of ecological barriers and the role of isolation in allowing ecosystems to become so diverse. Chapter 3 takes a look at the development of transportation technology and international commerce—the driving force behind biological invasions worldwide.

In a series of pairs of chapters, Part Two examines the ecological consequences of and the human responses to invasions. The first chapter in each pair is a case study; the chapter that follows it explores the larger themes that frame the issues raised in the case study. The first pair—Chapters 4 and 5—documents the zebra mussel invasion in the rivers of Ohio, and examines the characteristics of effective invaders and the ecological consequences of invasions. Chapters 6 and 7 show the impacts of a century of invasive pests and diseases in the forests of the eastern United States, and lay out the procedures and policies that can help prevent more invasions. Chapters 8 and 9 evaluate the recovery of an island off the southern California coast ten years after the eradication of feral sheep, and describe the range of chemical and mechanical control methods available to deal with invasive species problems. Chapters 10 and 11 detail the history of a decades-long campaign to use natural enemies to control a major weed on the Great Plains, and outline the safe and appropriate use of biological control against this and other nonnative pests.

In Part Three, the focus shifts to what people can do about biological invasions. Chapter 12 is a story about the restoration of both ecological and human history in an urban natural area. Chapter 13 returns to Hawai'i to evaluate an ambitious campaign to improve public awareness of invasive species threats. Finally, Chapter 15 offers a few ways that individuals can help reduce the impacts of invasive nonnative species, and list a few books and Web sites that readers might find useful.

# PART ONE

# The Globalization of Nature

## The Causes and Consequences of Biological Invasions

# CHAPTER 1

# From Endemic to Generic

## Feral Pigs and the Destruction
## of Hawaii's Native Forests

The hallways of Los Angeles International Airport are lined with posters—offering glimpses of faraway places—Bangkok, Kathmandu, Paris—each billed as uniquely exotic. I was rushing to catch an 8:30 flight to Honolulu when one of the destination posters caught my eye. It showed a lush ravine lined with flowering trees and shrubs, with a series of knife-edged ridges in the distance. Hawai'i. An exotic destination if there ever was one.

Or, more properly, an *alien* destination. It is unlikely that anything in that poster was native to the islands. The showiest of the flowering trees were African tuliptrees. The understory was mostly Indian kahili ginger. I was headed for a place that was largely no longer itself.

When the first Polynesian explorers arrived in the Hawaiian Islands about sixteen hundred years ago, upwards of 90 percent of the species they encountered were found nowhere else. This high level of endemism—the quality of belonging or being unique to a particular place—arguably made the islands the most biologically distinct place on the planet. These first Hawaiians converted some of the native lowland ecosystems to agriculture and caused the extinction of a number of species endemic to those habitats. However, because these people brought relatively few nonnative species with them on their 2,400-mile canoe voyage, the damage they did to Hawaii's native ecosystems was mostly limited to the lowland areas where they lived and farmed. (One notable exception is that a number of upland species of flightless birds were driven to extinction at least in part through overhunting.) Their impacts were not substantially different from those of any other agricultural people who settled a place where farming did not pre-

7

MAUI

East Maui Road

Haleakala
elev.
10,023

HANA

Kipahulu Valley

Haleakala
National Park

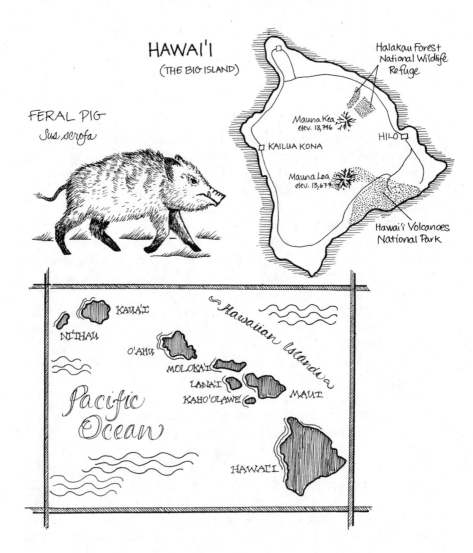

PROTECTED AREAS

HAWAI'I

(THE BIG ISLAND)

Halakau Forest
National Wildlife
Refuge

FERAL PIG
Sus scrofa

Mauna Kea
elev. 13,746

HILO

KAILUA KONA

Mauna Loa
elev. 13,671

Hawai'i Volcanoes
National Park

KAUA'I

NI'IHAU

Hawaiian Islands

O'AHU

MOLOKA'I

LANA'I

MAUI

KAHO'OLAWE

Pacific
Ocean

HAWAI'I

viously dominate the landscape: they cleared land, hunted animals, and gathered plants. Beyond the margins of their fields, Hawai'i remained largely Hawaiian.

The wholesale transformation of the undeveloped Hawaiian landscape gained speed only in the nineteenth century, when European explorers and settlers began to introduce one species after another to the islands in an attempt to make what they saw as a strangely depauperate and seemingly fragile environment a little more bountiful and robust. The response of the ecosystem to this constant influx of nonnative species made apparent some of the fundamental differences between an isolated island ecosystem such as Hawai'i and a continental system elsewhere in the world. The islands' ecological history created an inherent vulnerability to invasion by introduced species. Subsequent human-induced degradation further promoted the eventual devastation of the islands' native communities by invaders. Species by species and acre by acre, native Hawai'i has been disappearing since the moment it first appeared on a map.

When I got off the plane in Honolulu, the smell of the breeze across the open-air concourse was instantly familiar. It reminded me of Panama, where I'd lived for two years as a Peace Corps volunteer. Both the feel and the substance of the Hawaiian landscape were generically tropical. Apart from the mountains in the distance, there was nothing within miles to distinguish this from any other coastal city in any other tropical country in the world. Much of Hawai'i—in fact, almost all of the inhabited portion of the state—is now as generic as a strip mall.

Below 2,000 feet elevation, nonnative species overwhelmingly dominate the Hawaiian landscape, and even at higher elevations many native species are losing ground. Overall, nearly 50 percent of the species found in the wild on the islands are introduced. A large and growing portion of the state—in 1999, about half of Hawaii's land area—is no longer a native system that has been invaded by alien species; it is an alien system with a few native hangers-on. This process of transformation is creeping steadily upslope, each year moving a little higher into the remaining native forests.

Though many introduced species are implicated in the conversion of native systems to alien ones, there is a strong consensus among scientists and land managers that feral pigs, more than any other invader, are now responsible for initiating and driving this process further into the state's remaining intact native forests. As direct agents of forest destruction, pigs root up the soil, trample native vegetation, and eat sensitive native species. But their indirect impacts are perhaps even more significant. By creating openings and rooting the soil, pigs open up forests to invasion by disturbance-adapted nonnative species—many of which are transported by the pigs themselves, as seeds in their droppings or in their coats.

I have come to the islands to document the many impacts of pigs on native forests, both directly and through the self-promoting complex of alien species that accompanies them. But I am also here to investigate the cultural and political implications of feral pig management, for this is an issue where the biological and the political are inseparable. Though the technology and the resources to control pigs are readily available, the issue of eradication—even in dedicated natural areas—is highly contentious. Whether any native forest still exists in Hawai'i a century from now depends to a significant degree on how feral pigs are managed over the next couple of decades.

## A Refuge for Native Hawai'i

It takes at least a full day to hike into Kipahulu Valley. The mountains and cliffs that ring this valley in the highlands of eastern Maui comprise some of the most difficult terrain on the islands, and no public trails lead in from the outside world. But very few people have ever gone in on foot anyway. Kipahulu is a closed scientific reserve, and those with permission to enter generally do so via helicopter. Such rugged terrain has kept out more than just hikers and explorers. In this section of Haleakala National Park is some of the highest-quality forest remaining in Hawai'i—a forest where virtually everything is native, and almost all the historically recorded species still survive. This valley is the kind of place that has earned Hawai'i a reputation among conservationists as one of the most beautiful and most unusual places on Earth.

But no place in Hawai'i—not even Kipahulu—is entirely safe from invasion by nonnative species. The valley is crisscrossed by fences and laced with snares, the tools of a pig eradication campaign that began in the 1980s and that keeps the valley pig-free today. Even so, nearly a half century of pig presence prior to the eradication campaign allowed a number of nonnative plants to make inroads into the native vegetation, and the most frequent visitors to the valley today are Park Service weed control crews. It would now take, at most, only a decade or two of inattention even for Kipahulu, one of Hawaii's most pristine places, to fall into irreversible degradation.

Kipahulu's first systematic exploration came in 1945, when a pair of Park Service rangers ventured over from the Haleakala Crater to investigate the possibility of incorporating the valley into Haleakala National Park (at that time known as Hawai'i National Park). Chief Ranger Gunnar Fagerlund and Ranger-in-Charge Frank Hjort hacked through thickets of *uluhe* fern and

The Kipahulu Valley is one of the most isolated and pristine ecosystems left in the Hawaiian Islands. (Steve Anderson)

worked their way around cliffs on a five-day hike down to the coast. The forest was so dense and the terrain so rugged that even these experienced outdoorsmen covered only a mile or two a day. They reported seeing pigs just over the ridge from Haleakala Crater, but pig sign disappeared as they moved down into the valley. The valley was essentially free of alien plants all the way down to about 1,800 feet elevation. Fagerlund and Hjort described the valley as an ideal example of untouched "virgin wilderness" and strongly recommended that it be protected.

For the next twenty-two years, Kipahulu disappeared from the map. An occasional hiker might have ventured over the rim of the valley from Haleakala Crater and pig hunters probably worked the lower elevations, but there is little chance that anyone set foot in the heart of the valley until 1967, when a scientific expedition sponsored by The Nature Conservancy spent a month documenting the valley's biological richness. The 1967 expedition found that, unfortunately, while human attention was turned elsewhere, nonnative animals had been moving in from above and below. From the 7,350-foot-high valley rim down to about 6,300 feet elevation, the landscape showed serious pig and goat damage, and alien plant species were establishing and spreading wherever rooting and browsing were most severe. At the bottom of the valley, the upper limit of heavy pig activity had moved from 1,800 feet to 2,500 feet. And perhaps most significant, pigs were present even at the valley's once-pristine core. Though the expedition's report noted that pig disturbance in the central valley "appears to be minimal as yet," it emphasized in its recommendations that "immediate steps should be taken to reduce or eliminate the wild pig population. . . . Their damage to the vegetation is insidious but serious, and has in addition led to the establishment and spread of exotic plants in the valley."

The 1967 report provided the impetus for The Nature Conservancy to purchase a portion of the valley and transfer it that same year to Haleakala National Park. (The state of Hawai'i transferred ownership of the remainder of the valley to the Park Service in 1974.) The Park Service's initial management strategy was simply to designate the area as a closed scientific reserve and allow no public access whatsoever. However, reports from park rangers of serious weed infestations along the route of the main 1967 trail prompted the Park Service to send a small expedition into the valley in 1976 to ascertain whether human disturbance had opened up the area to invasion by alien plants. The 1976 expedition found that weeds had indeed spread significantly in some areas and that lower section of the 1967 trail—from about 3,500 feet elevation downward—was dramatically degraded. But the human impact of the earlier expedition was not responsible for the difference at lower elevations. Pig sign was everywhere in the area below 3,500 feet, and what had been mostly intact forest understory in 1967 was now a vast pig wallow. "It is impossible to determine what the effects of the 1967 expedition were in this area," the report stated, for "any effects have long since been masked by the effects of pigs."

The members of the 1976 expedition were worried not only by the presence of pigs but also by the spread of the strawberry guava tree into the valley from below. They also highlighted the possibility (confirmed in recent years) that "pigs may play a role in its spread, both in carrying seeds and in removing competing vegetation." Though earlier reports on the status of the valley had mentioned such a connection, the 1976 report was the first to note just how serious a threat the synergistic relationship between pigs and nonnative plants was to the integrity—and even the continued existence—of the native forest ecosystem of Kipahulu. Pigs were not only moving guava and other weeds up into the forest; the weeds were moving the pigs as well, for new infestations of guava trees served as higher-elevation food resources for pigs. "The current situation in the [lower] part of the valley . . . should probably be described as an emergency," the report concluded. In the strongest of language, it then urged that pigs be eliminated completely,

Nonnative feral pigs have devastated many of Hawaii's remaining natural areas. (Jack Jeffrey Photography)

before the pig-driven process of forest transformation degraded the valley beyond hope of recovery.

No action was taken, however, until after a 1982 study assessed pig population levels quantitatively and made explicit the consequences of failing to remove the pigs. Among other likely impacts of continued pig presence, the study predicted that strawberry guava stands would continue to replace native forest, the pool of alien species would increase, and water quality would decline. These concrete data persuaded Congress to provide funding for eliminating pigs, and as the 1982 study neared completion, plans for an eradication campaign began to take shape.

At the same time, the Park Service launched an interdisciplinary research project to document ecological changes in the valley as pigs were eliminated from the ecosystem. The valley was divided into three management units, and all but the lowest reach of the valley (the area already dominated by aliens) was fenced in with hog-proof wire. Permanent transects and study plots were established in the middle and upper reaches of the valley to monitor density and impacts of pigs, as well as the relative proportions of native versus nonnative vegetation as eradication progressed.

Pre-eradication population densities in the valley were 6.0 animals per square kilometer in the upper (more isolated) unit and 14.3 per square kilometer in the lower unit—both significantly lower than in many other natural areas. The Park Service team maintained an average of nearly 2,000 snares in the field for the forty-five months of the control program, ultimately removing 53 pigs from the upper unit and 175 from the middle unit. By January of 1989, pig populations in both units had dropped to zero.

Four years after the last pig was killed, the Park Service team revisited two of the most heavily damaged study sites (one in the upper unit and one in the lower unit) of those that had been evaluated for pig impacts before control was initiated. In the upper unit site, 60 percent of the forest floor was bare ground and 40 percent was in native vegetation just before pig control. Four years after pig eradication, the forest floor in the upper unit was completely covered in native plants. In the middle unit site, 7 percent of the forest floor was bare ground, 3 percent was covered in native species, and 90 percent in alien grasses (all of which are significantly more tolerant of pig rooting than are native species) before the pig control program was initiated. Four years after pig eradication, 70 percent of the forest floor in the middle unit was covered in native species and 30 percent in alien grasses.

According to Steve Anderson, Haleakala National Park's Natural Resources Program Manager, the difference between areas that showed complete recovery of native vegetation and those that did not was probably related directly to the intensity and duration of pig impacts. In the upper unit, pig invasion was recent enough and of low enough intensity that there remained rhizomes of native *pohole* ferns in the soil. Though pigs had introduced nonnative Hilo grass even into these most remote sites, the fern rhizomes in the soil allowed native vegetation to resprout and overtop the grass soon after pigs were eradicated. In the lower unit, however, pigs had been present long enough and at high enough densities that significant areas were entirely devoid of native vegetation—even roots and rhizomes—when pig eradication began. As a consequence, nonnative species introduced by pigs were able to establish themselves in persistent pockets throughout the forest.

While the upper unit once again enjoys almost pure native cover, the fate of the lower unit remains uncertain. The difference in recovery between the upper and lower sites suggests that the postinvasion window of opportunity in which full recovery is possible may be quite short and that even relatively low pig densities can have significant effects on native vegetation. In fact, Steve Anderson is convinced that in the long run, there is no such thing as an acceptable pig population density in native Hawaiian forest, for even at very low population levels, pigs will decimate sensitive native plant populations, create disturbed areas in the forest, and bring in invasive weeds. "If you're managing for a full complement of native species," he said, "the only acceptable level is zero."

Much of Kipahulu was cleared of pigs within the window of opportunity that allows for full recovery of native vegetation. But many of Hawaii's other native forests have long since passed the point where the more sensitive native species are likely to recover to anything resembling preinvasion levels. For example, ongoing pig presence at the Big Island's Hakalau Forest National Wildlife Refuge may be driving to extinction the very native bird species and rare plants that the refuge was established to protect. Though this forest is also considered one of Hawaii's most ecologically valuable, the degradation it continues to suffer even under conservation management may prove irreversible. Whether Hakalau survives as a native forest community depends largely on how quickly the refuge is able to follow Haleakala's model and eliminate pigs.

## Pigs and Birds in the Hakalau Forest

Jack Jeffrey doesn't like to spend much time in the office. He is a wildlife biologist, and there isn't much of biological interest in the vicinity of the U.S. Fish and Wildlife Service office in Hilo. When we met in the parking lot at 7:00 in the morning to head up to Hakalau Forest National Wildlife Refuge—the 33,000-acre tract of rain forest on the windward side of the Big Island that Jack is charged to protect from pigs, rats, weeds, and a horde of other threats—he didn't waste any time on administrative details. As soon as the rest of the expedition arrived, we jumped in his battered green Fish and Wildlife Service truck and headed up the road toward the refuge.

Hilo is at the center of what was once a major sugar-producing region. The abandoned canefields that stretch for miles out of town are now overgrown with a tangle of nonnative trees and grasses. It is only above the upper reaches of the former agricultural lands—around 2,500 feet elevation—that small native 'ohi'a trees begin to dominate the landscape. Past the old canefields, the road climbs through sparse, lava-strewn 'ohi'a forest for about half an hour, but at about 6,000 feet, native vegetation gives way once again to agricultural lands. The mamane-koa forest that once blanketed the shoulders of Mauna Kea at these upper elevations was destroyed many years ago by cattle, goats, and sheep, and has been maintained as pasture ever since.

We turned off the main road and cut across the pasture to the northwest. It was an hour-long drive on a bumpy gravel track to the Hakalau field station. As we bounced and swerved along the road, Jack generally managed to keep at least one hand on the wheel as he related the natural history of the forest and railed against the pigs that have invaded Hakalau. The two have been intimately linked since long before the refuge was established in 1985, and the continued presence of feral pigs is the single largest threat to the native species—even to the entire natural community—that the refuge is mandated to preserve.

"What is this refuge here for?" Jack demanded. The answer came in the same breath. "It's here for preserving habitat for endangered species," he exclaimed, with a clear emphasis on "endangered." Under this definition, nonnative feral pigs clearly do not qualify as legitimate beneficiaries of refuge management. However, Jack is frustrated by the fact that pig control efforts are proceeding slowly and sporadically and have yet to reach large portions of the refuge. Among Jack's primary responsibilities as refuge biologist (and what I was going to help him with today) are periodic surveys of pig activity at Hakalau. It is a part of his job he fervently wishes he no longer had to do.

We pulled into the field station around midmorning. From the station down to the forest edge, the dirt road wound between young koa trees that volunteers have planted by the thousands across the abandoned pastures that comprise the top portion of the refuge. Getting the trees to survive and establish was initially rather difficult, for open pasture is a much harsher environment than the small forest gaps to which koa seedlings are adapted. With a little experimentation, though, Jack and his colleagues developed a set of techniques that served as a surrogate for the protection that would have been offered by surrounding trees, and survivorship increased dramatically. Now birds are beginning to use the trees, and a few native plants—mostly tree ferns—have taken root in their shade. Over time, the planted koa trees will form a "nurse" environment that will allow the native ecosystem to move upslope into the old pastures. "Plant it, and they will come," Jack said. The strategy appears to be working. Some of the more common native birds have already been sighted in the young koa groves.

The immediate goal of reforestation is simply to begin the restoration of native vegetation on the degraded upper reaches of the refuge. Over the longer term, though, its purpose is more specific and in a sense far more urgent than overall ecosystem restoration: to reconnect the koa-'ohi'a forest in the refuge with the remnant forest of mamane trees further upslope. The goal of such a project is to reestablish an unbroken corridor from mid- to high-elevation forests, for it now appears that the survival of many of the refuge's birds may in the long run depend on access to higher-elevation habitat. While all of Hawaii's common endemic forest birds and five of the island's seven endangered forest birds are found on the refuge, today almost none of the rarer species are found below 4,500 feet. At lower elevations, there are nonnative mosquitoes that carry introduced diseases—avian malaria and avian pox—against which native birds have virtually no resistance. Below this invisible line, bird mortality increases sharply, and the endangered species are the first to go.

But all of this replanting may be too late for the endangered birds. *Culex quinquefasciatus*, the species of mosquito found on the islands, is distributed across the tropical, subtropical, and warm temperate regions of the world, and there is no biological reason why it can't survive at much higher elevations. Though surveys have so far found few adult mosquitoes in forests above 4,200 feet at Hakalau, incidence of disease is definitely moving upslope. As late as 1977, the Hawai'i creeper was found almost down to 2,600 feet, but in 1995 it was found only above 4,500 feet. Scientists studying avian disease believe that although breeding populations of mosquitoes are now established only at lower elevations, the species is such an efficient vector of avian disease that the small numbers of mosquitoes blown to higher elevations by onshore winds may be enough to infect and kill significant numbers of native birds well above the elevation where mosquitoes currently breed. The replanted koa forest might not establish itself quickly enough to provide mature habitat in time for the birds—especially given the pressures that are driving disease upslope through the refuge.

The most worrisome of these pressures are the habitat changes caused by feral pigs. We parked the truck at what is now the upwardmost edge of the forest and began our hike down into pig territory. For the first hundred yards or so, we followed a fence between a unit with many pigs and one with relatively few. The difference between the two was obvious: On the side of the fence with high pig density, pig sign was everywhere, and there was as much bare ground as grass (most of it nonnative). But on the other side, we had to work our way through and around and under downed logs and tree ferns and thickets of native shrubs. Unlike the highly disturbed tract on the other side of the fence, the matrix of this forest was still largely composed of native species.

Nonetheless, damage in the unit with fewer pigs was everywhere—the signs were simply harder to see. We stopped for lunch in a small clearing left by a downed tree fern-a sure sign of pig presence. The starchy cores of tree fern trunks are a preferred food for pigs, and toppled ferns are a common sight in this forest. Their hollowed trunks are the source of another subtle but catastrophic consequence of pig presence: Once a tree fern's core is eaten out, the trough that remains fills with water and becomes a perfect location

Well-maintained pig-proof fences are an essential component of the management of Hawaiian natural areas. (Jack Jeffrey Photography)

for mosquitoes to lay eggs. Though there is sometimes enough standing water in streambeds at Hakalau for mosquitoes to breed, pigs increase the availability of standing water in the forest. Given the efficiency of even very low densities of mosquitoes at transmitting avian disease, the habitat modifications that pigs cause are at least partially responsible for the fact that there are no endangered native birds below 4,000 feet at Hakalau.

According to Dennis LaPointe, a scientist who studies avian disease at the U.S. Geological Survey Pacific Island Ecosystems Research Center's Kilauea Field Station in Hawai'i Volcanoes National Park, the Hakalau Refuge is courting disaster by focusing its efforts on the top of the refuge. "Pig management should be from the lower boundary up, not from the upper boundary down," he said. No one can be sure how quickly disease will continue to move upslope, but pig disturbance in the lower reaches of the refuge can only make the invasion spread faster. Jack Jeffrey was just as emphatic about the urgency of the pig problem. "Time is of the essence with these birds, and we have to do something soon," he said. "If we don't have the backing of the public and the money to do control work, I see us still killing pigs twenty years from now. And by that time the birds will survive only at the upper reaches of the refuge—if they are still around at all."

Jim Jacobi, another biologist stationed at the Kilauea Field Station,

This native Hawaiian *apapane* is being bitten by a nonnative *Culex* mosquito, a species that carries diseases deadly to Hawaii's native birds. (Jack Jeffrey Photography)

agreed. He emphasized that there would be a significant lag time between the elimination of pigs and the disappearance of the breeding grounds they have created, for the cavities created by pig-toppled tree fern trunks may persist for years. "This is a desperate situation," he said, "and we've got to deal with these problems *now*."

In addition, there may be very little time left to save the critically endangered plants on the Hakalau Refuge. On the way back up to the truck, Jack pointed out a native lily growing high in the cleft of a tree. "Where there are no pigs, that plant is common on the ground," he said. Many of the endangered plants on the refuge are in trouble specifically because pigs find them particularly attractive. One such plant—*Cyanea shipmanii*, a palmlike understory species—is represented in the wild by only five known mature plants (and a hundred or so seedlings in Hakalau's plant nursery). Constant disturbance and predation by pigs prevents effective reproduction, and it is likely that the natural seed banks for these species are tremendously depleted. Chronic low-level pig impacts are not as dramatic as massive invasion, but for the most sensitive species, the long-term consequences can be equally catastrophic.

After a certain point—and at Hakalau that point may have already been passed—spontaneous recovery of such species becomes unlikely, even after many years of release from pig pressure. For example, the Waikimoi area on Maui was noted in the 1920s for the tremendous numbers of lobeliads (a group of native understory plants) that grew there. When pigs invaded the area some time around midcentury, the lobeliads were probably among the first to suffer significant damage. In 1983, The Nature Conservancy purchased Waikimoi as a nature preserve and promptly began a pig eradication program. But although pigs have been eliminated from most management units and are scarce in the remainder, some species of lobeliads are still rare, as are the native honeycreeper birds that rely on their nectar as a food source. At Hakalau, it is likely that many sensitive plant and animal species have already suffered similar long-term damage.

According to Dick Wass, the refuge manager for Hakalau, it will take at least twelve more years to clear the entire refuge of pigs. "We don't actually have a target date for complete eradication," he said. "We don't want to do this all at once." In the mid-1990s, control efforts were stalled for a year as the refuge wrote a "Feral Ungulate Management Plan" and prepared an Environmental Impact Statement for the plan. In 1998, control was just getting under way once again, but it will be years before lower remote sections of the refuge are even fenced, let alone cleared of pigs. The Fish and Wildlife Service has not developed nearly as much momentum or capacity in feral ungulate control in Hawai'i as has the Park Service, and its lack of experience shows in its reluctance to pursue with single-minded dedication its mandate to protect native species and ecosystems. "I'm really concerned," Jim Jacobi said, "because I don't think Fish and Wildlife has the staff or the money to deal with these problems." The future does not look bright for the birds and plants that Hakalau was established to protect.

The long-term consequences of continued pig impacts at Hakalau have the potential to extend well beyond the loss of native birds and understory plants. Scientific research in Hawai'i is just beginning to expand its emphasis from endangered species preservation to whole-ecosystem conservation, and a major focus of this work will be the impacts of the decline or loss of individual species on overall ecosystem integrity. Although work is in the early stages and no concrete evidence yet exists, it is conceivable that the loss of some key pollinator of a major native tree species could have cascading effects throughout the forest ecosystem. Perhaps most worrisome is that the ecosystemwide effects of the loss of this kind of key species would not be felt until long after it was too late to do anything about it. "If you've frayed the fabric too much," said Maui forester Bob Hobdy, "it falls apart." Where the point of such catastrophic change falls will vary with circumstances, but there is always a point beyond which recovery is not possible. Hakalau may still be a decade or two short of that point. But the longer pigs remain, the

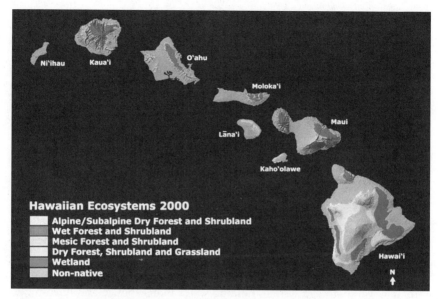

About half of the state of Hawai'i—the lightly shaded coastal areas on this map—are mostly or completely dominated by alien invasive species. (Hawai'i Natural Heritage Program)

more likely it will be that the forest will enter a long downward spiral of degradation from which there can be no return.

Many places in Hawai'i are already well past this catastrophic point. The Nature Conservancy's base map of Hawaii's remaining native communities leaves about half of the state blank. These are the places where most conservationists have given up, because the native ecosystem has essentially disappeared and any natives that remain are no more than isolated remnants. The conservation community has learned to let the few natives in these places go because, as Alan Holt of The Nature Conservancy put it, "No matter how much money you spent, you could not bring these places back." Places that are past the point of no return are of conservation interest only in that no one wants to see them grow any bigger. They are like a cancer that metastasizes, spitting out invaders into the still-healthy portions of the Hawaiian ecosystem. With dedication and consistent effort, the invaders might be contained indefinitely—but such places are themselves lost.

## "The Hawaiian Tropical Paradise"

We could have been driving through the poster I saw at L.A. International Airport. The narrow coast road along windward east Maui was lined with gorgeous tropical foliage and jammed with tourists who should have been keeping a closer eye on the road. The views from this road are the kind that have earned Hawai'i status as a "tropical paradise" in the minds of visitors and residents alike. But like the scene in the airport poster, this paradise is Hawaiian only in the eyes of those who created it. Almost all of what is native to this place is gone. These are the blank places on the map.

Bob Hobdy and I wound our way across one lush tree-lined ravine after another, with mountains rising from the road's edge and ocean reaching out to the horizon. Bob has seen many changes in the island's forests—most of them for the worse—in his years as a forester and a naturalist on the island of Maui. We took a drive along the windward Maui coast on a sunny January afternoon to discuss the impacts of a century of management on the island's forests.

Forestry in Hawai'i began not out of concern for the forests themselves, but rather for the vast quantities of fresh water they provide. In a land of steep slopes and no lakes, forest foliage and root-bound soil are the islands' principal water storage system, holding water like a sponge and releasing it in a steady flow. As early as the late 1800s, agricultural interests began advocating forest protection, and many large private plantations and ranches set aside forested land above their fields in order to ensure an adequate water supply. The territorial government followed suit in 1903 with the creation of a Forest Reserve System, with the intention of maintaining permanent forest cover on territorial lands that were not used for other purposes. By 1939, more than a million acres of reserve land set had been aside, about 150,000 acres of which were on Maui.

While much of the land designated as reserves was still in native forest, significant portions had already been denuded by loggers and ranchers or by feral animals at the time they were incorporated into the system. As a consequence, tree planting was nearly as important an activity of the Division of Forestry as was protection of existing forests. The vast majority of the trees planted, however, were nonnative. In 1908, a massive unexplained dieback of native forest on flat-topped ridges in Maui caused great alarm throughout the islands and fueled the drive to plant nonnative tree species. Prevailing opinion in the early years of the twentieth century was that decline of native forests was inevitable and that replacement with fast-growing, aggressive introduced species was the only way to maintain forest cover. As a consequence, Hawaii's forests did not simply succumb to self-spreading introduced trees; in many cases, they were actively replaced by foresters who believed that just about anything was better than native forest that died inexplicably and left nothing but shrubs and ferns to hold the soil.

But Bob Hobdy is not convinced that the native forests were intrinsically that fragile. While research in the 1970s and 1980s demonstrated that 'ohi'a dieback is in many cases a natural, cyclical phenomenon, Bob suspects that this particular forest dieback may also have coincided with the first large-scale invasion of these native 'ohi'a forests by feral pigs. In that case, the real cause of the "Maui forest disease" may have been a catastrophic alteration of basic ecosystem dynamics. In fact, Bob said, chances are good that rooting and compaction by large numbers of pigs could have made the wet soil on flat-topped ridges anoxic (greatly deficient in oxygen), killing tree roots and inhibiting the establishment of seedlings. While large feral mammals were considered vermin and pig control was a part of forest reserve management programs at the time, early foresters did not have the resources to eliminate pigs from the vast areas of forest that they had invaded. Pigs pushed the 'ohi'a into a steep decline, and the loss of native forest became a self-fulfilling prophecy as nonnative trees were brought in.

Bob has witnessed the loss of many thousands of acres of native forest in his years on Maui. He pointed over the mountains toward a remote area on Maui's far east end that he knows well. In the 1970s, he had flown over what had been a near-pristine 'ohi'a forest in that area just after it had been invaded by pigs. "The effect was really catastrophic," Bob said. The pigs had turned the whole place into a mudfield, and today, the entire area is a sea of alien weeds. In this case, the process of conversion to an alien system needed no help from humans. Even though many of the overstory trees survived the initial invasion, the forest ecosystem is on its way out, for the weeds prevent the establishment of native tree seedlings. The remaining mature trees are little more than standing ghosts.

Pigs are implicated in the destruction of native wet forest more often than any other introduced species. What they leave behind is the kind of semifunctional hodge podge of alien species that we drove through for hours along windward east Maui. "That's the stunning thing here," Bob said. "Most places you go in the world, you can find at least some natives mixed in. But here the domination is complete." We pulled off to the side of the road so I could take a few pictures. I asked if there was anything native within view of where we stood. Bob spent a minute surveying the vegetation before he answered. "No," he said slowly. He shook his head as we got back in the truck. This native forest has not simply been altered, he explained. It has been replaced. This is an entirely new ecosystem.

And this is a system that continues to change as more invasive species are introduced. A few miles down the road, we stopped at the Waikamoi Nature Trail, a popular tourist spot known for its 2- and 3-foot-thick euca-

In some places, a second wave of even more aggressive invaders is replacing earlier non-native arrivals. This stand of Australian black wattle trees is being smothered under a carpet of South American banana poka vines. (Hawai'i Division of Forestry and Wildlife)

lyptus trees (an introduced species) that were planted in the 1930s to control erosion. As we walked through the grove, Bob pointed out that there are virtually no young eucalypts taking hold among the thickets of ardisia—a large introduced shrub—that forms the understory along the trail. Ardisia and a handful of other relatively recent introductions appear to be the next wave of dominant invaders in Hawai'i. Since the environment they have invaded—the alien-dominated lowland forest—was itself composed of aggressive nonnative species, the plants that comprise this second stage of invasion are tougher still. Most are short-statured, shade-tolerant plants that in their native habitats are either understory or early successional species. Here in Hawai'i, these invaders tend to form impenetrable 10- to 30-foot-high single-species stands under which almost nothing else will grow. The result is a tangle of undergrowth so aggressive that it allows no overstory. Lowland east Maui may be turning into a forest without trees.

Bob is not optimistic about the future of Hawaii's remaining native forests. "I believe that we have set in motion biological processes over which we have scant control," he said. There are dozens, perhaps hundreds, of introduced plants present in Hawai'i that are waiting for an opportunity to invade a favorable environment and spread explosively. Healthy native forests can resist invasion by all but a handful of these introduced species—but only so long as they are free from nonnative disturbance. Wherever pigs and other introduced mammals incur upon native forest, that forest will suffer drastic losses of sensitive native plants, invasion by alien weeds, and in some cases a drop in native bird populations. Inevitably, that forest will be pushed toward the kind of generic ecological pandemonium that is already the rule in much of what once was Hawai'i.

## The Making of a Vulnerable Ecosystem

Are Hawaiian ecosystems uniquely vulnerable to catastrophic change as a consequence of invasion by nonnative species? Do islands suffer from an intrinsic fragility unknown on the mainland? The evidence would seem to indicate that they do—or at least that Hawai'i does. For example, nonnative birds introduced to Hawai'i have successfully established at a much higher rate than is the rule for introductions on continents. While an average of only about 10 percent of continental introductions become naturalized, more than half of the birds introduced to Hawai'i by European settlers have breeding populations on the islands. And the fact that half of the state (including large stretches of uncultivated land) is dominated by nonnative species serves as evidence that introductions to Hawai'i have been remarkably successful.

The most commonly cited reasons for vulnerability to invasion in the Hawaiian ecosystem and in island ecosystems are ecological. For instance, many insular species are not as well defended against predation, disease, and herbivory as are their mainland counterparts, which means invaders are often presented with abundant, easily consumable food resources. (Jack Jeffrey said that from a nonnative feral animal's point of view, Hawaii's forests are "just like a salad bar.") The absence of many kinds of organisms in the native biota of most islands—few or no large predators or herbivores, for example, and few diseases and parasites—provides invaders with unoccupied niches in which to establish and frees them from the competition and predation that otherwise would keep them in check. As was noted in a recent analysis of Hawaii's predicament, "Isolated oceanic islands were predisposed to certain types of human-related invasions because of long isolation from the continual challenge of some of the selective forces that shape continental organisms."

But these sorts of intrinsic ecological factors are not the only reasons for Hawaii's vulnerability to invasion. Some kinds of vulnerability have been induced by the choices people have made about land use in an island ecosystem, and Hawaii's current state is as much a product of its history as of its biology. There are several ways in which human actions have served to exacerbate the intrinsic vulnerability of this isolated ecosystem.

First, Hawai'i was by no means a "pristine" environment when Europeans arrived and began to introduce large numbers of alien species. Polynesian settlers and their descendants had at one point farmed or burned almost all land below 2,000 feet elevation, and this kind of massive disturbance probably made these lands significantly more vulnerable to invasion than the undisturbed ecosystem would have been. In addition, Polynesian hunting pressure had already driven a number of bird species to extinction before European settlement, creating artificially open niches in bird communities that left them more vulnerable to invasion by introduced birds. In short, the fact that some Hawaiian ecosystems were significantly altered by humans for a thousand years before European-induced invasions began may have been just as responsible for the vulnerable condition of the Hawaiian biota as were the evolutionary effects of isolation.

Second, Europeans tended to discount the value and the hardiness of native island species, and as a consequence, the Hawaiian chain has been subjected to an inordinately large number of species introductions in historic times. State agencies and private individuals have brought in more than 80 species of game animals to provide sport in an environment that had almost nothing native that was worth shooting. Plant enthusiasts have imported showy ornamentals from across the tropics, and foresters have planted anything that might conceivably grow, together adding roughly 10,000 alien plant species to the islands' 1,000-species native flora. In the early 1900s, a group of Honolulu residents went so far as to form a club whose sole purpose was to import birds to Hawai'i, bringing in 52 nonnative songbirds over the course of several decades. Perhaps in part because European settlers saw Hawai'i as an empty place waiting to be filled, its native biota is outnumbered by attempted introductions to a degree not often seen in continental ecosystems.

Finally, existing invaders have begun to create opportunities for new invasion that have little to do with the inherent vulnerability to invasion of the original native ecosystem. For example, the spread of feral pigs into native forest has been driven at least in part by the range expansion of strawberry guava and other nonnative food sources such as introduced earthworms. There are dozens of such examples of ecological relationships between nonnative species that have eclipsed the constraints of the original system, where the rate of establishment of new nonnative species is increasingly independent of the characteristics of the native ecosystem. While

Hawaii's vulnerability has its origins in the islands' evolutionary ecology, the process of ecosystem transformation that threatens to destroy what remains of native Hawai'i has taken on a life of its own.

What this all means is that induced vulnerability to invasion may in the long run be at least as significant as intrinsic vulnerability in determining the pace and the path of ecosystem conversion to domination by alien species. As a consequence, continental ecosystems may under some circumstances become just as vulnerable to catastrophic transformation as are island ecosystems. Many different combinations of intrinsic and induced vulnerability can create an opening for the kind of invasion that can trigger a critical shift in some fundamental ecosystem process such as nutrient cycling, food webs, or disturbance regimes, thereby putting a broad spectrum of once-dominant native species at a general competitive disadvantage. This sets up a positive feedback cycle by which nonnative species rapidly take over both the dynamics and the composition of the system.

## The Mechanics of Pig Control

While the consequences of this kind of positive feedback are sometimes largely irreversible, in other contexts there is still much harm to be prevented and much restoration that can be done. In any ecosystem—insular or continental—often the most powerful means of preventing or even reversing ecosystem transformation by invasive alien species is to eliminate the forces that fundamentally alter ecosystem processes. In Hawai'i, this above all means eliminating pigs from native forests. Pigs change the rules of the game, forcing native species to compete in a system whose dynamics no longer match those species' evolutionary history. As Alan Holt of The Nature Conservancy of Hawai'i has argued, the Hawaiian ecosystem is durable and viable only if "the fundamental rules don't change too much." Trying to protect native species without first getting rid of pigs and other feral ungulates is like trying to push water uphill—you would do much better to realign the terrain first. Therefore, while combating specific plant invaders is often useful and sometimes quite necessary, weed control alone is not as effective as eliminating the engine that opens up the system and drives in the weeds to begin with.

The technologies for eradicating pigs are well developed. Hawai'i Volcanoes National Park on the Big Island was a pioneer in feral pig eradication, and in the 1980s, the park showed that eliminating pigs was possible even in dense forest and extremely rough terrain. The key to its success was single-minded determination in pursuit of its goal. Beginning in 1980, park personnel fenced off tracts of land, used professional hunters and trained dogs to bring the pig population level down, and then set snares and traps and hunted by helicopter throughout the area to kill the few remaining pigs. Most of the pigs were killed during the first six months of the program, but complete eradication took about three years and required the use of snares and traps.

The extra effort required to kill the last few pigs—and it was a major effort, since the ones that consistently manage to evade hunters are tremendously wary and elusive—was absolutely crucial. Merely keeping pig populations suppressed through sporadic public hunting would have done little for the preservation of the park's sensitive natural areas, for at least 70 percent of a population must be removed each year in order to have any effect on pig numbers. (Between 1930 and 1980, over 11,000 pigs were killed in the park mostly through sporadic hunting, with no appreciable effect on long-term pig numbers.) Even a professional hunting program would have been a waste of both time and money if it had failed make complete eradication its goal. This kind of thorough, systematic approach has been the model for every one of the natural areas in Hawai'i that has been successfully cleared of pigs.

Given the political will, it would be possible to eliminate pigs from most if not all remaining natural areas on the islands within a decade or two. But while some progress is being made, eradication is not happening nearly as fast as it could. The most immediate reason for slow progress is lack of funds. However, the real impediment to eradication is not financial, but social. Pig hunting is of great cultural significance to a small but highly vocal segment of the state's population (some native Hawaiians, others descendants of more recent settlers). In addition, some animal rights groups have in recent years made pig snaring one of their key issues. So long as a significant portion of the population remains sympathetic to the notion that pigs are legitimate residents of native forests or believes that pig presence is preferable to snaring, Hawaii's native species will continue to disappear as pigs degrade the state's remaining natural areas beyond recognition.

## Pigs and People

Pigs have been in Hawai'i for almost as long as people have been there. The second wave of Polynesian settlers probably brought pigs with them by canoe when they arrived around the twelfth or thirteenth century. However, the pig that the Polynesians introduced is not the same animal as the one that is destroying forests today. The modern feral pig is descended largely from

European pigs that escaped or were released from hog farming operations over the last few hundred years. While Polynesian pigs were generally rather small and rarely ventured far from the coastal villages where they were raised, their European cousins are much larger and more fecund and move readily into remote areas.

Unfortunately, very few people make any distinction between these two very different animals. They have transferred the cultural significance of the Polynesian pigs onto the mostly European hybrid that has replaced them—the only pig most hunters have ever known. The confusion is compounded by the fact that rural culture in Hawai'i has largely shifted from a subsistence to a cash economy. What is now largely sport is still suffused with the language of need, creating a strong emotional argument that plays well in the popular press. Articles with titles like "Oahu Pig Hunters an Endangered Species" and "As Preservation Efforts Rise, Hunters Fear Own Extinction" are common in Hawai'i newspapers, and any threatened loss of hunting ground is met with vigorous protest.

Many hunting advocates feel that they have already given up too much to a cause whose benefits they do not value. George Martin, a pig hunter on the Big Island, doesn't believe it can be proven that pigs are destroying the forests. The problem, he says, is that scientists have too low a threshold for what constitutes "destruction." He concedes that there is a role for fencing and eradication, but only of very small plots to protect representative samples of rare plants. Fencing pigs out of thousands of acres of forest, George contends, "is wrong." He believes that pigs in fact do a service to the forest by rooting out nonnative species—a contention any scientist will hotly dispute—and that the forest can adapt to pig presence.

The consequences of this culture are manifested in the patchwork of conservation areas and pig management areas that continue to vie for position on the Hawaiian landscape. For example, the two major units of Hakalau Forest National Wildlife Refuge are divided by a tract of state land that is managed specifically for a sustained yield of pigs. Statewide, there is one federal program that provides Hawai'i with funds to manage nonnative game animals and another that pays for fencing and eradication of these same animals, often on adjacent tracts of land. Even on land dedicated to native ecosystem preservation, pig control is sometimes too hesitant to be effective because pig hunters with little knowledge of ecology are allowed to influence decisions about ecological conservation. Until land managers feel they have a local as well as a national mandate to eradicate pigs, only the most creative and persistent among them will be able to muster the resources and the backing to get the job done.

Pig eradication is further complicated by an animal rights campaign that targeted The Nature Conservancy and the state of Hawai'i throughout the early 1990s for their use of snares. People for the Ethical Treatment of Animals (PeTA) launched a letter-writing campaign, picketed The Nature Company and other major corporate donors to the Conservancy, promoted antisnaring legislation in the Hawai'i legislature, and vandalized several Conservancy preserves in Hawai'i. Both the Conservancy and the state were forced to divert significant resources to countering the misinformation that was spread as part of the campaign. Ironically, some hunters joined forces at least temporarily with the animal rights campaign to ban snares. Their combined efforts set back pig control significantly in several natural areas at a time when there was no time to lose.

At the same time, the antisnaring campaign forced the conservation community to articulate the rationale for snaring in particular and for pig control in general, and to find ways to create a larger community of support for native forest protection. The Nature Conservancy took the lead in improving public appreciation for native ecosystems and public understanding of what it takes to protect them, but many other agencies and individuals did outreach work as well. The key was to make abundantly clear what was at stake: the survival of Hawaii's remaining native forests and of many of the species that comprise them.

For most people, the rationale for pig eradication makes sense once they learn something about the ecological consequences of inaction. Virginia Mead, a member of both The Nature Conservancy and PeTA, had initially been opposed to the use of snares, believing that there ought to be alternatives to snaring even in the most remote and rugged sites. "I went to the PeTA-TNC meeting pushing for an end to the snares," she explained. "I came out understanding that removing the snares before a viable alternative is implemented dooms the native animals to a cruel death and the extinction of their species." And Eddie Oliveira, an east Maui native and lifelong pig hunter, recognizes the importance of pig eradication as well. "To preserve our native forests, we need fencing to keep the pigs out," he said. His appreciation for native forests comes from a lifetime spent in the mountains, and the fact that he is on good terms with both the native forest and his human neighbors has done much to advance the cause of conservation in his east Maui community. Hawai'i needs many more like him.

The last time I was in Hawai'i, I stopped in at a small barbershop in the village of Mountainview on the Big Island to get a haircut. The shop was in the front room of a house on the only street in town. I sat down in the chair, and the woman who ran the shop asked me what brought me to town. I told her I was writing a book on the impacts of nonnative species on natural areas and was headed up to Hawai'i Volcanoes National Park to talk with a couple of biologists about pigs. "Oh, those pigs do a lot of damage," she said. And as she cut my hair, she talked about growing up on the island and about the native birds she used to see when she was young.

More than fences, more than snares, more than weed crews, Hawaii's forests depend on people like her. As much as it is driven by ecology, conservation is ultimately a human enterprise. The fight to protect Hawaii's native diversity from pigs and weeds must be won in barbershops and living rooms before it can hope to succeed in the remote valleys and mountains.

# CHAPTER 2

# Private Worlds

## *The Relationship between Ecological Isolation and Biodiversity*

Hawaii's presettlement native flora and fauna would not have made a biodiversity "hot spots" list by species counts alone. Hawai'i has only 1,131 native species of vascular plants. By comparison, Massachusetts—never thought of as a hotbed of diversity—is home to 1,700 native vascular plants. The primitive Hawaiian ecosystem was not particularly well endowed in other species groups either. It was overall a rather depauperate community.

About sixteen hundred years ago, Polynesian colonists landed on Hawai'i and initiated a transformation of the islands' biota. They cleared lowland areas for agriculture and drove about fifty bird species to extinction through hunting and habitat destruction. But the Polynesians brought several new species with them, including groups such as terrestrial vertebrates (in this case, pigs and rats) that previously had been wholly absent. Even though some native species disappeared with the arrival of the Polynesians, on the whole the ecological changes they wrought caused an increase in local biological diversity, at least in terms of the breadth of species types represented.

In 1778, the first European sailors came to the islands, and shortly thereafter, species introductions began to increase dramatically. Plants were introduced as crops and ornamentals, and mammals and birds were released for sport and pest control. Some insects invaded as stowaways, and others were released as biological control agents. By the late twentieth century, the number of alien plants and animals growing on the islands exceeded 10,000 species, and nearly 1,000 of these nonnative species were established in the wild. Even taking human-caused extinctions into account, there are now more than twice as many species present on the Hawaiian

Islands as there were before humans arrived. And more new species are arriving nearly every day.

In the more-is-better school of diversity, Hawai'i is in great shape. The reality, though, is that Hawai'i is experiencing a major conservation crisis. The reason is simple: For the purposes of raw diversity counts, a species is a species is a species, but for conservation purposes, endemic diversity is critical. Most of the species brought into Hawai'i are common the world over, and their addition to Hawai'i adds nothing to global biological diversity. Hawaii's loss of endemic species, on the other hand, reduces the world's overall biological richness.

Although humans have added many alien species to the native Hawaiian flora and fauna, Hawai'i nevertheless has one of the highest rates of endemism of any place on earth. Among its native insects, about 99 percent are endemic; among flowering plants, 94 percent; and among terrestrial mollusks such as land snails, 99 percent. The islands' native biota as a whole is over 95 percent endemic. The native Hawaiian ecosystem is as close to unique as anything on earth. As a consequence, its contribution to global diversity is significant despite its relatively low number of species. Hawaii's high endemic diversity is the direct result of one simple fact: Hawai'i is—or was—the most isolated ecosystem on earth.

Isolation and time are the key forces driving the creation of unique species. What isolation creates depends on the circumstances, and Hawai'i is an extreme example of the effects of multiple scales of isolation—from that of the island chain as a whole to that of individual valleys divided by high ridges—on the diversity of an ecosystem. But this process is at work everywhere on earth, and the ways it acts upon natural systems are the same in any context. Understanding how isolation creates and maintains diversity is just as important to conservation in New England or the northern Rockies as it is to conservation in Hawai'i, for the loss of isolation is one of the major factors disrupting native communities worldwide.

## Biodiversity and the Importance of Place

Mapping the geographic distribution of earth's species is a necessary part of understanding the world's biodiversity, for a living creature's place is a critical part of what defines and sustains it. But biogeographic knowledge is often limited to the very familiar. Most people know that lions are found in Africa but not in South America. Their current presence in India and former presence in Europe, though, is much less widely known. Kangaroos are of course from Australia, but most people would be hard-pressed to tell you whether kangaroos were also native to New Zealand. (They aren't.) Only a handful of people are likely to be familiar with the consequences for the Mediterranean of digging the Suez Canal: an invasion of Red Sea fish and invertebrates. Yet the world needs biogeographic knowledge to guide important decisions at every level, from local landscaping to federal importation regulations. As a highly mobile species, humans need to understand the role of place—and the *separation* of places—in shaping the biotic world.

The global distribution of biodiversity tends to follow a few big patterns. These patterns map out the associations of species that have evolved in one another's presence and as such give an indication of the consequences of disrupting these associations through careless ecosystem mixing.

1. *Biodiversity is mostly tropical.* Terrestrial and some marine ecosystems tend to increase in diversity from the poles to the equator.
2. *Old habitats are rich habitats.* The longer it has been (on a geological timescale) since the last ecosystem-wide catastrophic disturbance, the more diverse a system is likely to be.
3. *Divided landscapes create more species.* Island archipelagoes and other kinds of segmented habitats are fertile places for geographic speciation.
4. *Habitat diversity promotes species diversity.* Increased variation in topography, soil type, and other physical ecosystem attributes generally leads to increased biological diversity.
5. *Species diversity is self-reinforcing.* Increasing biotic diversity stimulates further increases in biotic diversity—for example, more species of plants allow for the existence of more species of herbivorous insects.
6. *Isolated places are more likely to have unique species.* Habitats that are extremely isolated and very old tend to have more unique species than less isolated areas (though they don't necessarily have more species overall).
7. *Physical constraints promote diversity in some habitats.* Moderate levels of physical disturbance and moderate nutrient deficiencies may promote local diversity by preventing dominance by a few highly competitive species.

BIODIVERSITY IS MOSTLY TROPICAL. Fund-raising campaigns and TV news stories with powerful images of the Amazon on fire have made tropical rain forests and biodiversity synonymous for most Americans and Europeans. While moist tropical forests are not the home of all biological richness, in many ways they are evolution's favorite child, and their fame for diversity is well deserved. Consider hummingbirds. New England is home to only one species: the ruby-throated hummingbird. Farther north, in Labrador, gardens lack even this single species. To the south, the Florida-to-Arizona latitudinal belt has fifteen species, but the center of hummingbird richness is northern South America—home to ninety-eight species of hummingbirds.

This pattern of biodiversity—in which the number of species increases toward the equator—repeats itself for many kinds of organisms: bats, lizards, birds, flowering plants, and insects, among others. Exceptions do occur; for instance, some groups prosper in harsh places but cannot compete where conditions are less severe. Lichens can endure cold, dry conditions, but are overwhelmed on good soils in milder climates by faster-growing grasses and herbs. For lichens, the icy north or arid deserts are centers of diversity precisely because their competitors cannot thrive there. In general, though, evolution appears to have gone to more elaborate lengths in the tropics than anywhere else.

OLD HABITATS ARE RICH HABITATS. No matter what the geographical context, evolution takes time. Whether diversity increases slowly or rapidly depends on many factors, but large-scale catastrophic disturbance—for example, the advance and retreat of continental glaciers—will always set the clock back. Consequently, some of the most diverse ecosystems on earth are those that have remained relatively undisturbed the longest. The fynbos, an extremely old area in the Cape region of South Africa, has some 6,000 species of native plants, most of them found nowhere else on earth. New York State, which until 15,000 years ago was under mile-deep ice, is roughly the same size as the fynbos but has far fewer species, most of which are found throughout eastern North America. Old lakes, such as Lake Baikal in Siberia and Lake Malawi in East Africa (the former 25 to 30 million years old, the latter 5 to 9 million years old) are rich in locally evolved species as well. Lake Malawi has more than 600 unique fish species, and Lake Baikal is home to almost 240 species of unique crustaceans. In contrast, young lakes like the North American Great Lakes (all less than 15,000 years old) have few locally evolved species and are relatively poor in overall species diversity.

DIVIDED LANDSCAPES CREATE MORE SPECIES. Fragmented systems produce more species by virtue of an abundance of physical barriers. Island archipelagoes are by definition divided habitats and are better at fostering bursts of speciation than are single islands. The twenty-seven Galápagos Islands off the Pacific coast of Peru support fourteen finch species, all derived from a single ancestral species that colonized the archipelago. In contrast, Cocos Island, a single island off the Pacific coast of Costa Rica, supports only one finch species.

Other physical features may also divide habitats into multiple sectors, each of which is equivalent ecologically but isolated physically (at least to some species). Such features also allow separate evolution in each local area. On the Hawaiian island of Oahu, high, knife-edged ridges divide the island into a series of isolated valleys, and in each valley, a distinct race or species of *Achatinella* land snail has evolved. Snail populations have remained separate because ridgetops are unsuitable as snail habitat, and separation over millenia has allowed for speciation.

HABITAT DIVERSITY PROMOTES SPECIES DIVERSITY. Mountainous locations provide diverse habitats ranging from wet to dry and warm at sea level to cold on the mountaintops. Such regions are a mosaic of habitats in which evolution is likely to generate a diverse cast of species, each adapted to a particular set of conditions. For example, the high islands of Hawai'i show tremendous variation in rainfall and altitude from one locality to another, and therefore support more species than do the low islands in the chain.

SPECIES DIVERSITY IS SELF-REINFORCING. In 1733, writer Jonathan Swift observed that "a flea hath smaller fleas that on him prey." His point is well taken, for while previously established species may be effective competitors that exclude some new species, plants and animals are themselves resources available to be exploited by newly colonizing or newly evolving species. High plant diversity, for example, promotes high herbivorous insect diversity, which in turn supports a high diversity of wasps and other creatures that parasitize the herbivores. A rain forest with tens of thousands of plant species will of necessity be richer in insects than will a desert with only a few hundred species of plants.

ISOLATED PLACES ARE MORE LIKELY TO HAVE UNIQUE SPECIES. As mentioned earlier, species found only in one place are known as endemic. How big a place the species is native to doesn't matter. The Bermuda cedar is found only on the tiny Atlantic island group of Bermuda. Kangaroos are found only on the continent of Australia. Both are endemic, but on vastly different physical scales.

Endemic species form where both isolation and time abound. Isolated arctic islands that have only recently emerged from beneath the ice rarely have endemic species for lack of time. Coastal islands are often too close to the mainland for isolation of any but the least mobile species. Endemic species form most readily on old oceanic islands, in old lakes, in isolated river systems, or in isolated patches of distinctive habitat such as caves, hot springs, or odd soil types. The dodo of Mauritius, the unique cichlid fishes of Lake Malawi, and the blind fish and crawdads of some Appalachian limestone caves are all examples of endemic species formed in isolation.

PHYSICAL CONSTRAINTS PROMOTE DIVERSITY. In the absence of disturbance, the composition of a community naturally changes over time from species that are good colonists to species that are good competitors. Highly competitive species often dominate habitats in more stable environments and exclude many of the early colonists that thrive only on disturbed sites. Consequently, moderate levels of disturbance can maintain a mosaic of site conditions that allows many types of species to thrive over a broader area. Similarly, patches of lower-fertility soils often allow a larger mixture of species to persist because such soils are not favorable to the growth of the more vigorous plant species.

For example, fire promotes diversity in North American oak savanna habitats by killing all but a few oak seedlings, allowing a wide variety of grasses and wildflowers to thrive that otherwise would be outcompeted by oaks. When European settlers began suppressing fires in oak savannas, diversity plummeted as oaks and other tree species sprouted in large numbers and shaded out the understory plants that depended on fire to maintain the conditions they needed to survive.

## Isolation as a Biological Multiplier

Every one of the patterns described above demonstrates one or more of the effects of isolation on the character and composition of ecosystems. But isolation affects global diversity differently in each situation. Sometimes isolation functions as a niche duplicator, allowing related but distinct species to fill similar ecological roles in separate places. In other cases, isolation enhances ecological opportunities by restricting the numbers (and kinds) of competitors that any given species must face. Isolation also permits parallel solutions to the same environmental challenges, allowing different taxonomic groups to evolve to fill the same ecological roles in separate locations. Finally, isolation puts limits on predation by narrowing the range of attackers a species has to cope with in order to survive.

DUPLICATING NICHES. In both island and continental ecosystems, ecological isolation allows for duplication of niches by permitting multiple biological "solutions" to the same ecological "problem." For example, before Polynesians spread throughout the Pacific, hundreds of closely related species of flightless rails (small wading birds with long beaks that often live in marshes) inhabited the islands of the region. Nearly all were single-island endemics, derived from a single closely related species of normal flighted rail. Each was unique to the one place it lived. This great variety of species existed only because the islands they inhabited were separated by the nearly insurmountable barrier of large stretches of open ocean. These interisland barriers of ocean effectively expanded the "wading-bird-with-long-beak" niche into multiple copies of itself, allowing separate species to evolve on each island. Hypothetically, if all the islands could have been pulled together into a single large landmass, many of these distinct species would have disappeared because the best competitor (or at most a few species) would have preempted the resources over the whole geographic area. But since the niche was divided into many independent parts, a great variety of species came into being. What was one opportunity in ecological theory became many opportunities in geographical reality.

PROVIDING ECOLOGICAL OPPORTUNITIES. In some ecosystems, isolation also provides opportunities for evolution to fill niches in unusually creative ways. Isolated habitats (either real islands or ecological ones) often show significant gaps in the kinds of colonist species that reach the habitat. A common result of such gaps is that lineages "stretch" over time, branching out into ways of living from which they normally would be excluded by superior competitors. For instance, one of the finches on the Galápagos evolved to occupy the niche of woodpeckers through the behavioral adaptation of using a cactus spine to pry insects out of plant tissues—a niche it probably would have been unable to fill had a "real" woodpecker species been present.

In general, the fewer kinds of colonizing species present in an ecosystem and the longer their history of isolation, the more diverse the ecological repertoire of each group may become. On continents this sort of lane-shifting is rarer, because there often is not enough empty ecological "space" in the community due to the large number and kinds of species present. Occupants of "stretched niches" on islands may be quite seriously affected if the continental occupant of the same niche ever shows up. If a real woodpecker were ever introduced to the Galápagos Islands, for example, it would likely be more efficient than the false woodpecker finch and might outcompete it or even drive it to extinction.

MAKING PARALLEL EVOLUTION POSSIBLE. When a single lineage (such as the rails of the Pacific oceanic islands) colonizes a range of isolated locations and evolves into many separate but related species, the resulting forms look alike to varying degrees because of recent common ancestry. In other cases, unrelated organisms that live in widely separated locations but are adapted to similar habitats may also show many shared characteristics. Desert cacti in the Americas and some African euphorbs that live in dry areas look much alike, but these similarities are common design features evolved independently by separate lineages to cope with the same environmental constraints—in this case, the need to conserve water.

Isolation at a grand scale, such as between the biogeographic provinces of the Americas and Africa, allows independent solutions by separate lineages to a single problem. This is an example not of duplication of niches, but rather of convergence within a niche, whereby species that share very little taxonomic history come to resemble one another because of adaptation to similar ecological conditions. One might predict that true cacti, if introduced to Africa, might outcompete their euphorb counterparts (or vice versa) because such loss of isolation would pit two desert-adapted lineages against each other—a form of competition that the isolation between Africa and the Americas prevented during the evolution of these two groups.

LIMITING THE RANGE OF ATTACKERS. Another very important function of isolation is to define and limit the range of antagonists that a species must be able to withstand. Any successful plant must survive its herbivores and pathogens. Animal species must survive their predators and parasites. Species develop traits that help them escape antagonism, suppress the level of antagonism, or rebound from its effects. All such traits, however, are contextual, and selection will act only with reference to the antagonists that have played a meaningful role in the evolutionary history of the species.

Species can be expected to have some tolerance for antagonists with whom they have long been in contact. However, new associations can be devastating. Human populations in isolated areas have frequently been decimated when they were first exposed to a new disease. Many indigenous North American peoples, for example, died in large numbers when contact with Europeans first exposed them to smallpox. Similarly, plant and animal species will be shaped by evolution to tolerate only those antagonists with which they have been in contact. For example, island plants may lose defenses against grazing mammals if they have existed for many generations on islands lacking herbivorous mammals. Plant species on continents may lack defenses against pathogens or insects that invade from other biogeographic provinces. The American elm and the American chestnut were both decimated after contact with exotic pathogens. Similarly, virtual extermination of the forest birds on the island of Guam followed the invasion of the brown tree snake because these birds had no ecological or evolutionary familiarity with such a predator.

## The Separation of Worlds

Diversification plays out differently at the extremes of ecological isolation. At one endpoint are the strange and beautiful evolutionary creations that come of the niche-stretching diversification that occurs on oceanic islands, such as the two-ton moas of New Zealand and the shimmering silversword plants of Hawai'i. At the other end of the spectrum is the orderly within-niche elaboration of continental creatures no less beautiful, but more neatly within the realm of the expected—from the brilliant tanager songbirds in a Costa Rican rain forest to the hundreds of species of delicately differentiated pearly mussels in North American rivers and lakes.

But the extremes of "island" and "continent" are best understood as models, not as real ecosystems. In reality, every ecosystem sits at many points along this spectrum at once. Every continental ecosystem contains isolated pockets with unfilled niches; and no island, no matter how isolated, is free from the constraining forces of competition. Isolation occurs on many scales—from what continent a species lives on to what other species it eats—and the power of isolation as a biological multiplier is made all the greater by the layers of complexity these different scales create.

At the largest scale of isolation—known as macroscale isolation—terrestrial habitats on earth are divided into the continents. The flora and fauna of separate continents are largely distinct, and the continents are the largest divisions (called biogeographic provinces) into which living forms are grouped. These divisions are stable only to a degree, because over geological time, landmasses move over the earth's crust, combining and splitting apart in patterns that have arisen and disappeared many times over the earth's history. Continents are floating arks—each bearing its particular cargo of animals and plants—and interchange happens only when the continents come together or when a plant or animal manages to cross the barriers between the continents. Under natural conditions, both of these events are extremely rare.

One of the world's grander natural experiments in species mixing provides an example of how great a role macroscale isolation can play in the maintenance of global diversity. When the formerly separate regions of North and South America were joined some 3 million years ago by the rise of the Panamanian Isthmus, the new land bridge permitted groups such as nonflying mammals to move easily between the continents for the first time.

Numerous species exchanges occurred in both directions. Marsupials such as the common opossum moved north, while placental mammals such as llamas, cougars, peccaries, and jaguars migrated south.

On first thought, one might imagine that both regions would gain by the exchange. Instead, competition thinned the ranks, eliminating about 24 percent of the mammal families present in the combined faunas of the two provinces before they joined. This reduced the pool from fifty families before the land bridge formed to only thirty-eight after the exchange of faunas. Separateness had supported more diversity because some combinations of competition and antagonism did not occur when potential competitors lived in isolation from each other.

And what of the oceans? As a terrestrial species, humans look at continents and see an enormous variety of environments: deserts, swamps, mountains, fertile valleys, icefields. The oceans are to most people little more than water. But a closer look reveals the richness of ocean biogeographic provinces as well, which are far older than the biogeographic provinces on land. Oceans are both connected and subdivided by planet-sweeping currents such as the Gulf Stream, creating provinces called "gyres" with distinctive communities of organisms. The gyres in the Pacific, Atlantic, and Indian Oceans and the smaller seas such as the Mediterranean, Red, and Caspian once held separate and largely distinct sets of species, with very little natural interchange between them. Much of this distinctiveness has been lost to the building of sea-level canals, the movement of stowaways in association with shipping, and deliberate interocean introductions.

When it comes to maximizing diversification, however, groups of semi-isolated ecosystems provide the ideal balance between isolation and interchange. Interchange allows new species to arrive, though at low rates; isolation gives these species time to evolve into new forms that are distinct from the ancestral line. This middle ground is known as mesoscale isolation, and it occurs on both real and ecological islands.

Island chains provide isolation at two levels: continent-to-island and interisland. The effectiveness of distance as an isolating mechanism is to a certain degree modified by additional factors such as changing sea levels, the direction of prevailing winds, seasonal storm systems, and ocean currents, all of which may act as dispersal mechanisms for colonists between islands or from mainland sources. Long chains of islands are particularly profligate, because while movement from one island to the next might be relatively easy, separation by two or three hops generally constitutes a highly effective barrier to interchange. The hundreds of thousands of islands across the globe are home to a significant portion of the world's biological diversity.

At this scale of isolation, it becomes clear what "island" means in an ecological sense: not simply a bit of land surrounded by water, but any discrete area of a given kind of habitat that is embedded in an environment inhospitable and impassable to a large portion of the inhabitants of that habitat. The key is not distance between islands per se, but rather the idea of "inhospitable," whatever that may entail in a particular ecological context. Half a mile of desert sand is just as effective an isolating mechanism for fish in a desert pool as is a hundred miles of open ocean for bats on a small tropical island. Crossings happen, but not all that often.

Even within continents or islands, the world is still patchy. Sand dunes may occur along the coast in a broken chain, interrupted by occasional miles of forested headlands. Bogs may be scattered sparsely among vast areas of boreal forest. If such patches of distinctive habitat are sufficiently separated one from another, they act as ecological islands. Many landscapes are mosaics of small patches of many kinds of habitat. The small barriers created by such mosaics result in what is known as microscale isolation. The particular impact of such patchiness on evolution depends on the mobility of a species. Mobile species will be little affected, but species with low dispersal abilities may exist only as disconnected populations in disparate patches. For bog-inhabiting birds, the patchiness of bogs is not a barrier; for bog orchids, each local population may exist in nearly complete reproductive isolation.

Barriers do not have to be physical—or even immediately visible—to be effective. Even in the absence of physical barriers, other processes may operate that isolate local populations of a species and prevent them from exchanging genes. The most powerful such influence is a close association of an exploiter with its host. For example, populations of an herbivorous insect species that feed on different kinds of plants may divide into separate species, each associated with its own particular host plant.

This highly specific sort of separation can be called bioscale isolation. It is most common when there is a behavioral link between a species' food source and its mating habits. For example, the apple maggot is a native North American fly whose original host was hawthorn, a wild relative of apples native to North America. After apple was introduced to North America, the fly adopted apple as an additional host. The insect population, however, found it difficult to be optimally adapted to both plant species. Because breeding in this group of flies occurs on the fruit, flies that by chance were especially responsive to the odor and form images of apple fruits mated preferentially among themselves, separating themselves in the process from flies breeding on hawthorn. While they are still considered a single species, populations of this fly on each of these two hosts appear to be diverging and are viewed by many entomologists as a case of speciation in progress. Such behaviorally based isolation can permit subpopulations to maintain their genetic separateness long enough for genetic differences to become large, thereby isolating them from each other reproductively—at which point a new species has arisen.

At every scale, isolation is a large part of what makes a place unique. Ecological barriers create and maintain the particular character of a place. The breakdown of barriers leads to loss of many unique species and in the long run tends to homogenize communities across the globe.

## When Worlds Collide

In an article published in the journal *American Scientist*, four biologists who study alien species issues calculated the hypothetical impact on global mammalian diversity of bringing all the world's land area into a single supercontinent—an effective total loss of macroscale isolation. Based on the number of species per square mile on each of the continents, they projected that such a landmass would support about 2,000 species of mammals. However, with the continents separate, the earth actually has about 4,200 species of mammals. The authors concluded that "the complete breakdown of biogeo-

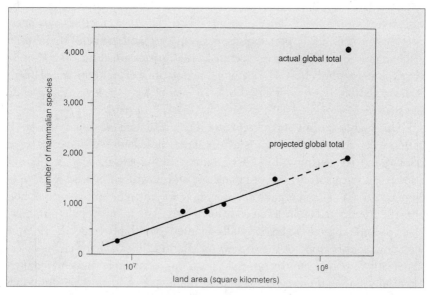

This graph illustrates one hypothetical consequence of a complete breakdown of ecological barriers: extinction of half the world's mammal species. (Courtesy *American Scientist*)

graphic barriers might result in the eventual extinction of more than half of the earth's mammalian species." While differences between faunal groups make it risky to project from this mammalian model to other kinds of organisms such as insects or plants, analogous patterns of loss may hold true for many forms of life. Directly or indirectly, the breakdown of barriers and the loss of isolation lead to the loss of species.

Biological invasions are nothing new. No place on earth has ever been totally isolated, and, on a geological timescale, the world is the richer for the natural exchanges between ecosystems. The arrival of colonists provides much of the raw material for evolution. What humans have changed is the rate of invasion, and as the authors of the above-mentioned article note, this change "is so large as to represent a difference in kind rather than degree." What once was a catalyst for evolutionary invention has become an overwhelming force for ecological destruction. The increase in the rate of species movement is of a magnitude that only a technological species such as humans could cause. As a consequence, it is without precedent in the history of the earth.

It is impossible to determine what constitutes a "natural" invasion rate in general terms. The rate of invasion of an ecosystem depends on how isolated any particular location is with respect to some species of potential colonists, and varies both from place to place and across different kinds of organisms. However, one can compare invasion rates before and after human settlement on oceanic islands to estimate the magnitude of the human-caused increase in invasion rates. Because oceanic islands were among the most isolated ecosystems on earth until the rise of *Homo sapiens*, they were subject to relatively low natural rates of invasion. Consequently, the increase in invasion rates on islands after human settlement provides a measure of how much humans have changed the rate of species movement around the globe.

Since the Hawaiian Islands are the world's most isolated ecosystem, their invasion rate before human settlement must have been among the lowest in the world. The origins and size of their native biota, their geological history, and approximately how many introduced species they have are all relatively well known. From this information, it is possible to arrive at a rough calculation of the increase in invasion rate for Hawai'i.

The biota of the Hawaiian Islands is believed to derive from about 1,000 ancestral species, and the ages of the currently extant high islands range from 500,000 to 5 million years. However, the process of island formation and destruction has been going on continuously for about 70 million years, and not all colonizing species are likely to have succeeded in leaving descendants on today's islands. The present biota of 1,000 founding lineages thus reflects a minimum number of colonization events over a period of somewhere between the age of the chain—70 million years—and the age of the oldest current islands—5 million years—giving estimates ranging from one species every 70,000 years to one every 5,000 years.

Since humans first arrived on the Hawaiian Islands, well over 1,000 alien species of plants and animals have become established in the wild on one or more islands. This has occurred in less than 2,000 years, for a rate of one species every 8 months—a rate from 2,500 to 35,000 times greater than the presettlement rate. This is probably a highly conservative estimate, and given that the vast majority of these human-caused introductions have occurred in the last 200 years, the present rate is higher still. What's more, the 10,000 to 12,000 species present in the state but not yet established in the wild are not included in this figure. The actual rate of introductions is probably significantly higher, and it increases every year.

As noted earlier, invasion rates in continental ecosystems can be described only in absolute terms. About 1,500 species of arthropods are known to have invaded the United States since 1800—around 7 per year. Tens of thousands of species of nonnative plants are currently in commercial trade. A recent book on naturalized vertebrates records the successful human-assisted invasions of 113 species of mammals, 109 species of birds, 10 reptiles or amphibians, and 67 fish in North America alone, many of which have had a significant effect on native plants and animals. Humans have certainly caused invasion rates to increase dramatically on continents as well as islands; it simply is not possible to determine exactly by how much.

The impacts of biological invasions on continents are in some respects quite different from their impacts on islands. Biological invasions have for several reasons been the direct cause of many fewer extinctions on continents than on islands. First, continental species generally have evolved in the presence of a wider range of pressures from predators, herbivores, competitors, and so on. This means that a continental species is, all other things being equal, more likely than an island species to possess defenses against whatever threats new invaders might pose. Second, even if a continental species with a large range cannot withstand the pressures brought to bear upon it by an invader and dies out in most places, it may persist in isolated pockets (known as refugia) where the invader is less successful. For example, eelgrass is a widespread marine plant along the eastern coast of North America that came in contact early in this century with an invading pathogen. The exotic pathogen was fatal to the plant, which was exterminated over nearly all of its range. Eelgrass survived only in a few areas of low salinity near rivers that discharged large quantities of fresh water into bays or estuaries because in these areas the pathogen did not thrive. From these limited refugia, surviving plants later recolonized much of the species' original range.

Although invasive species directly cause fewer extinctions on continents than on islands, the pressures they bring to bear on specific native species and the changes they make to invaded ecosystems put many species under constant stress. As a result, native continental species may become more vulnerable to extinction from other causes. Directly or indirectly, the biological invasions that come with increased species movement can cause fundamental, self-magnifying shifts in the species composition and function of native ecosystems, be they insular or continental.

In place after place around the world, the loss of isolation is replacing a tremendous diversity of small, ecologically rich worlds with a limited set of large, more homogenous ones. The consequences for global diversity are enormous. As the world economy pushes toward globalization, commerce and culture are dragging natural communities along with them. Even within the conservation community, the consequences of this loss of isolation often receive less attention than they deserve. But even with all that has been lost, so much more is still at risk.

At their root, nonnative species invasions are as much a cultural problem as an ecological one. Chapter 3 presents a historical perspective on the social and economic causes of this breakdown by tracing the global spread of *Homo sapiens*, and in particular the role of human transportation technology in the breakdown of ecological isolation.

CHAPTER 3

# The Great Mixers

## *Transportation Technology and the*
## *Spread of Invasive Species*

Like all other species, humans evolved in one place. Africa is the only part of the world to which *Homo sapiens* is truly native. It wasn't until about 200,000 years ago that the first humans walked out of Africa and began to spread around the globe.

By 100,000 years ago, humans were present in the Middle East, and by 35,000 to 40,000 years ago, they had reached Europe, east Asia, and Australia. The first humans arrived in the Americas about 11,000 to 12,000 years ago. Islands that had to be reached by long-distance ocean travel were the last to be colonized: Fiji and Tonga in the western Pacific were colonized only 4,000 to 2,000 years ago; the Marquesas Islands, Society Islands, the Hawaiian Islands, and Easter Island in the mid- and eastern Pacific about 1,500 years ago; and New Zealand, Madagascar, and Cuba only about 1,000 years ago.

These early journeys were made mostly on foot by small groups of people who had to carry everything on their backs. Consequently, they took little with them besides their clothes, weapons, household tools, crops, domestic animals, and perhaps a few hitchhiking parasites and weed seeds. At this stage of human history, relatively few other species traveled with humans, because cargo capacity was too limited and travel conditions were too harsh. The only major effect of the human diaspora was to seed the world with people.

As people invaded new lands—places where humans were themselves a nonnative species—they precipitated fundamental ecological change, especially on islands and in other isolated natural communities. For example, excavations of fossilized bird bones on the island of Ua Huka in eastern Polynesia show that on this island alone, at least twenty-one of the forty species of native birds once present on the island went extinct before Europeans ever set foot on the island, almost certainly as a result of first human colonization. A similar pattern of extinctions holds true for virtually all the islands invaded by the Polynesians, including the Hawaiian chain.

The marks of human impact are harder to see on continents than on islands, but in most cases the consequences of the arrival of humans were no less severe. In Australia, nearly nine out of ten genera of terrestrial megafauna (animals weighing 90 pounds or more) went extinct shortly after the first human invasion of the continent, as did about three-quarters of those in both North and South America. Recent evidence suggests that the timing of these extinctions was no coincidence; these animals were evolutionarily unprepared for such hunting pressure, and most were unable to adapt to human presence. It is interesting to note that, by contrast, the only continent that has not yet suffered massive loss of large native animals is Africa—the only place where *Homo sapiens* and large animals evolved in each other's presence.

While humans are clearly the most catastrophically invasive species of them all, the impacts of humans per se is beyond the scope of this book. The focus here is on the role of the human race in transporting other species around the globe. The history of transportation technology is fundamentally defined by *more*—more speed, more cargo, more distance, more often. For all but the last half century or so of human existence, how *much* more was constrained by the limited ability of technology to meet the challenges of the physical world. But in the age of globalization, all that holds back species movement is how much human societies choose not to do. For the first time in history, it is laws and good judgment—not cargo space or speed—that serve as constraints on species translocations.

## Living Locally in the Age of Foot Travel

For all but the last two centuries of the 200,000 years our species has been around, land travel has proceeded at more or less the same pace worldwide: 3 miles an hour, the walking speed of a person or a horse over rough terrain. This made most people's lives local affairs. If you had to walk somewhere and get back the next day, your radius of travel was under 20 miles. If you had to carry your own load, you were limited to about 50 pounds. With a pack horse, cargo capacity increased somewhat, but people still carried little more than the essentials.

Consequently, a late-seventeenth-century Massachusetts farmer taking butter to market moved no faster than the first humans did when they walked out of Africa. The reason for the lack of difference over this enormous period is simple: Most roads (if they existed at all) were little more than rutted tracks, unsuitable for anything more sophisticated than a horse or at best a horse- or ox-drawn wagon. With the exception of the road system built during the Roman Empire, no infrastructure had ever existed that could support rapid land transportation; thus, there was no incentive to invent anything faster than one's own two feet.

Domestication of animals for hauling cargo and the invention of wheeled vehicles some three to five thousand years ago dramatically increased the quantity of cargo that could be moved from one place to another. By 1600, large wagons pulled by eight to ten horses were the dominant means of shipping goods overland in Europe. Such wagons could move loads of up to 3 tons, making possible the movement of some species—mostly livestock, crop plants, and (inadvertently) weed seeds—across the continental landscape. However, livestock and wagons did not substantially increase the speed of transport. Goods still moved a maximum of about 20 or 30 miles a day, and as a consequence, few kinds of cargo were moved very far.

The essential limiting factor on overland species transport, then, was not cargo capacity, but lack of infrastructure to support rapid travel. In 1760, the 400-mile trip from London to Edinburgh took sixteen days—still a pace of about 3 miles an hour. While horse teams and wagons increased the loads that could be transported, they frequently got stuck or turned over because of poor road conditions, which kept average speed to a walking pace. Not until after 1800 did men such as John McAdam in England successfully champion public construction of roads built to higher standards, with even surfaces and good drainage. McAdam's roads quadrupled the speed wagons and carriages could travel to 12 miles an hour, cutting the London-to-Edinburgh trip down to just four days.

A few species did in fact move across the landscape with human travelers in the age of foot travel. But the slowness of travel and the relatively limited weight and volume of cargo that could be moved restricted both the kinds of species people would attempt to move and the kinds of pest species that could successfully travel as stowaways in association with people or their cargo. To survive anything more than a trip to the next town over, a species needed to be hardy, compact, and lightweight. Agricultural plants probably constituted the vast majority of species that were moved long distances intentionally, for the high cost of transportation meant that only high-value goods—items like spices, luxury cloth, and important crops—

were objects of long-distance trade. Nursery stock from distant lands, live fish, raw logs, and just about everything else were too heavy or too perishable to be moved, intentionally or otherwise. Some weed seeds were probably moved unintentionally as contaminants in crop seeds or attached to clothes or fur, but only a few were able to hang on for more than a few stops on a long overland journey.

Most overland species movement in this period would have been more akin to gradual expansion of a species' range, rather than to the disjunct ocean-hopping invasions that often leave controlling natural enemies behind (see Chapter 5). An overland traveler to China or other distant parts of Asia might be a source of such disjunct invasions, since the only major ecological barriers that travelers crossed on a regular basis before the age of sail were the deserts of central Asia. However, only a species capable of making the journey in stages would have accompanied travelers along the entire length of the route. Some agricultural pest insects are thought to have made their way west along the Silk Road in this fashion, moving slowly from China to the Middle East over decades or centuries by hopping from one oasis town to the next across the deserts of central Asia and feeding on fruit or plants moved by travelers.

Since species movement in the age of foot travel probably amounted to little more than the slow human-aided dispersal of a few exceptionally hardy species of agricultural weeds and pests, the world's ecosystems were overwhelmingly dominated by native species until about seven hundred years ago. Human presence had at that point already caused serious degradation in a variety of ecosystems—the near complete destruction of forest cover and loss of topsoil in Greece in ancient times is one example—but even in such places, what was left was largely native. It was only with the beginning of the age of sail around 1300 that the lines between the major biogeographic provinces began to blur.

## Bridging the Seas in the Age of Sail

Unlike wagons, boats ply a natural highway that needs no preparation, and the oceans are open to all who are brave enough to ride them. But while daring seamen have been around for millennia, sailing technology was too primitive to allow for regular, reliable ocean transit until about seven hundred years ago. It wasn't until after a series of key technological innovations had been made that sailors—almost exclusively Europeans—began to traverse the seas regularly and in large numbers. This second wave of global human migration launched the next stage of species mixing, for the age of sail made possible the mixing of the biota not only within continents, but between them.

Sailing began long before the oceans were crossed. Phoenicians are known to have sailed from the area of present-day Lebanon and Israel to the Atlantic Ocean and down the west coast of Africa over two thousand years ago. Stories even suggest that one voyage commissioned by an Egyptian pharaoh started at the Red Sea and, over the course of three years, successfully circumnavigated the African continent. But the Phoenicians' boats (as well as those of the Romans and other early Mediterranean sailors) were primitive in design and difficult to sail. Such journeys were rare, for they depended more on luck and skill than on good equipment.

These early galley ships had one mast and a single square sail, which meant they could sail well only downwind. If contrary winds were blowing, there was little that sailors could do to keep their ships from being blown onto reefs or threatening shores. Early sailors also lacked any reliable means of navigation on open seas. Once out of sight of land, sailors had nothing but the stars to guide them home, and without accurate timepieces and compasses, navigating by the shifting heavens was a daunting and uncertain prospect. At that time, almost all travel by sea was within a single biogeographic province: from one end of the Mediterranean to the other, or around the Baltic Sea, or from one South Pacific island to the next.

"Blue water sailors"—those willing to sail out of sight of land—had to wait for several major developments in sailing technology before they could make open ocean voyages with any reasonable expectation of being able to return home. They received an important tool when Crusaders brought back the compass (a Chinese invention) from the Holy Land. Together with maps, the compass allowed sailors to break away from coastal routes and set out across the open ocean without losing track of where home lay. But if they were to make long voyages safely and regularly, sailors also needed boats they could control in contrary winds. Single-masted boats with square sails hung perpendicular to the boat cannot sail upwind efficiently because the sail cannot be adjusted over a wide range of angles to the wind. Technical improvements that began to address this problem included rigging the sail fore-and-aft (that is, with the sail parallel to the length of the ship) and using several sails on several masts or a triangular sail at the front of the boat. Replacement of the ancient steering oar at the rear of the ship with a much larger and more easily controllable rudder board also helped provide better control. Once these critical innovations had been made around 1300, ship design evolved rapidly and sailors began to set their sights on more distant shores.

But even with compasses and improved ships, sailors still needed a working knowledge of the wind patterns that would carry them across the oceans—and back home. Between the 1330s and the 1520s, Portuguese and Spanish sailors discovered and mapped broad circles of wind that provided reliable routes across the Atlantic and back. A Portuguese sailor discovered the Canary Islands off the northwest coast of Africa in 1336 by following the northeast trade winds southwest from the Iberian Peninsula. Luck and insight eventually led the Portuguese to discover that the easiest way back was not to tack slowly and painfully against the wind up the African coast, but to sail west out to sea until they hit the westerlies, another wind that carried them northeast back to Europe. Other Portuguese and Spanish sailors mapped most of the rest of the great circulatory patterns of winds by the end of the sixteenth century, making possible the first controlled movement of large numbers of people and cargo back and forth between Eurasia and the Americas.

Several elements conspired to make Europe the place where this long series of inventions and discoveries—compasses and star charts, ships that could sail into the wind, and knowledge of wind patterns—came together to create a power that could sail around the world. Europe's physical position on the globe was as critical a factor in its transoceanic explorations as were its technologically advanced ships. Compared to the Pacific, the Atlantic was a manageable size to cross on a regular basis, and the Americas (once discovered) were a much more profitable destination than the scattering of islands in the Pacific or the emptiness south of the Indian subcontinent.

Perhaps even more important, European social and religious values tended to emphasize expansion and conquest, and the combination of missionary zeal, curiosity, and greed that characterized many European cultures both impelled their transoceanic voyages and fueled the innovation that made these voyages possible. China possessed many of these same basic technologies well before European sailors set out across the Atlantic, but Chinese cultural values tended to emphasize isolation over exploration. China had in fact sent a fleet of ships on an expedition that reached India almost a century before Columbus crossed the Atlantic, but on the fleet's return, the emperor decided that there was no reason to pursue exploration any further. From that point on, Chinese ships stayed home, and it was the fleets of Europe that came to dominate the world—and launch the process of global ecological homogenization.

The Polynesians also made enormously long ocean voyages well before the Europeans set sail, memorizing star maps and migrating in outrigger canoes across the Pacific to settle island after island across thousands of miles of open ocean. But since these voyages were highly infrequent and almost entirely one way, and since outrigger canoes were tiny and had little cargo space, they were poor vectors for spreading nonnative species. As a consequence, the Polynesians did little to break the isolation between the biota of separate continents. Their voyages were marvelous displays of sailing skill, but had little impact on the scope and scale of human transportation of other species.

When Europeans spread out of Europe after 1492, a torrent of species followed them across the globe. Once the best routes became known, the Atlantic could be crossed in three or four weeks. The cargo capacity of European sailing ships allowed seamen to bring a variety of seeds, horticultural and ornamental plants, and domestic animals with them, and provided a ready means of transportation for stowaways such as rats in the bilge and beetles and seeds in the soil used as ballast. The commercial routes between the Old World and the New initiated a flow of species across the oceans that has yet to stop.

Agricultural crops were generally among the first species to be introduced to an area during colonization. However, relatively few crop species have proven seriously invasive, at least in temperate climates. Settlers in North America, New Zealand, Australia, and South Africa brought most of their seeds and starter plants from Europe, but only a few species of clovers, pasture grasses, and plantation timber trees have become major pests of natural areas in the temperate zone. Colonists in tropical areas imported plants from other tropical parts of the world. The most famous voyage to move a tropical food crop was that of Captain Bligh of the HMS *Bounty*, who was charged with bringing thousands of young breadfruit trees from the South Seas to the West Indies, where they were to be planted to feed slaves. But many other tropical food plants commonly regarded as native to the Americas—citrus and banana, for example—are imports as well. Some translocated agricultural species have since became invasive, such as banana in Africa and South America, red quinine in the Galápagos, and strawberry guava in Hawai'i, among others.

Movement of live plants and plant products always involves the risk of moving insects with them. Many pest insects were rapidly moved around the world on plant materials sent between countries during and following colonization. For example, *Mayetiola destructor*, a small fly that is a major pest of wheat, invaded North America about 1779, probably in straw bedding used in troop transport ships when England sent Hessian troops to fight in the American Revolution. In fact, movement of crop pests was recognized as a major problem over a century ago, and some of the first quarantine laws in the United States and other countries were designed to counter this threat. However, relatively few of these crop pests have proven invasive in natural areas either.

Another important source of accidental introductions during the era of colonization was weed seed mixed in with crop seeds sent from Europe to colonies in temperate regions. Seeds to start hay meadows, for example, were harvested from European meadows and shipped as an undifferentiated mixture of species. Many European meadow plants—among them such common species as Queen Anne's lace, dandelion, and various wild mustards—are established in countries around the world. However, most (though not all) of these species invade only disturbed areas. They tend to be roadside plants and agricultural pests rather than invaders of natural areas.

Other early invaders arrived in association with the sand and dirt that sailing ships used as ballast. Since dirt was free and available everywhere, sailing crews departing Europe would take on enough dirt at the time of departure to balance the ship after the cargo was loaded. When the ship arrived in North America, the dirt was thrown out to make room for cargo being loaded for the return trip. The dirt was full of whatever insects and plant seeds happened to be found at ground level at the ship's point of origin, and the use of soil as ballast set up a flow of species between ports. Because young colonies exported a larger volume of goods than they imported, the net flow of dirt—and therefore of species—was from Europe to the colonies. This seemingly innocuous route provided a means of entry for many plants and animals, some of which, like the wetlands plant purple loosestrife, have become major pests of natural areas.

While agricultural crops have caused only limited problems as invasive species in natural areas, agricultural animals have been ecologically devastating to many of the ecosystems—particularly island ecosystems—to which they have been introduced. As they explored the world, European sailors made it a common practice to release pairs of livestock on oceanic islands as a means of providing food for sailors in the future. Wild cattle, pigs, goats, and sheep soon formed vast herds, for many island natural communities had no native mammals other than bats and thus livestock were not subject to predation. Also, the local plants often were palatable and relatively lacking in thorns or poisons to defend themselves against herbivorous mammals. Consequently, feral mammals multiplied until they consumed all the forage and starved. Degradation of the vegetation, plant extinctions, and soil erosion were commonly the result (for example, see Chapter 8). Some countries—especially New Zealand—have made a systematic effort to recover the native ecology of islands by instituting rodent, rabbit, and other verte-brate removal programs. Most islands, however, are still being ignored.

Other vertebrates came aboard ships and spread around the world as hitchhikers. Rats found European sailing ships quite commodious, and ships were typically infested with one or more species. Rats commonly jumped ship to colonize ports of call or were deposited on uninhabited islands when ships stopped for water or were wrecked on coastal reefs. Few invasive species have been more destructive of oceanic ground-nesting birds and lizards.

The invasion of nonnative species that was precipitated by European col-onization of North and South America, Australia, New Zealand, and oceanic islands has had profound ecological consequences for these places. But as effective as sailing ships were in breaking down the barriers between ecosystems, they could not do it all. At least on the continents, the impacts of the species that ships introduced were felt mostly around ports and in densely settled agricultural areas, for inland transportation remained slow and of limited capacity until the middle of the nineteenth century. Even sev-eral hundred years after the age of sail began, the interior of the newly colo-nized continents remained relatively untouched by nonnative species inva-sions.

The expense of ocean travel also restricted the kinds of cargo bound to the colonies largely to finished goods, machinery, and industrial products, while goods sent back to Europe were either processed products such as sugar, tobacco, and liquor, or high-value raw products such as furs, spices, sawn timbers, and dried fish. Shipment of vast quantities of low-value bulk goods such as raw logs of common species was not sufficiently profitable to be commonplace, which greatly reduced the risk of transporting pests and diseases that would attack native trees and other wild species. Finally, the several weeks it took a sailing ship to cross the seas meant that sensitive or perishable species—live plants, for instance, or freshwater fish—were diffi-cult to keep alive and were therefore only rarely transported. It took some-thing more than wind power to expand the pool of invaders and move the invasion inland.

## Saturating the Continents in the Age of Mechanized Travel

Sometime before 1850, a European wetland plant called purple loosestrife (see Chapter 11) jumped across the Atlantic to the New York–New Jersey coast in ballast soil in the hold of a wooden sailing ship and took root in North America. In the late 1980s, a Eurasian animal called the zebra mussel (see Chapter 4) came into the Great Lakes in the ballast water of cargo ships. Although the mussel invasion came nearly a century and a half after the arrival of purple loosestrife, the North American range of zebra mussel was within a decade nearly as large as that of purple loosestrife.

While biological differences between the two organisms are in part responsible for the zebra mussel's much more rapid spread, this dramatic acceleration of the tempo of invasion is even more the product of a continentwide transportation infrastructure that supports rapid, individualized travel. It took almost a hundred years for purple loosestrife to spread across the United States, moving slowly with floods and gardeners from one wetland to the next. Zebra mussels spread across the country as quickly as boats move from one place to another, traveling attached to barges or in the water wells of fishermen's boats that were put on trailers and moved from one watershed to another.

But the mechanization of land and water transportation did much more than simply increase the frequency and volume of routes by which nonnative species could move into a continent's interior. It also promoted the accidental introduction of a wider range of pests and diseases through increased trade in unfinished goods, and expanded the kinds of organisms that could be moved intentionally from one continent to another. The overall effect of mechanization was to push the barriers between ecosystems far lower than the sailing ships ever could have done.

Application of mechanical power to transportation started with boats on inland waters. Coal-powered steamboats were working the rivers in England by 1788, chugging along at 5 miles per hour. The major advantage of such power was the ability to move goods and passengers upstream. This proved particularly useful in countries like the United States and Brazil, where big rivers with strong currents existed as major potential conduits of trade. In North America, river trade before 1820 was generally a downstream affair. Rafts were built in a river's upstream reaches, loaded with goods, and floated downstream to a port. There the rafts were dismantled, for they could not return. Steam power allowed boats to push cargo and people—and therefore nonnative species, be they guests or hitchhikers—into the heart of a continent. These settlers brought many of the same species that the age of sail had brought to the continental margins well into the interior of the continents.

The next phase of continental penetration began in the mid-1800s with the construction of the railroads. Extensive rail systems developed before highway networks largely because a rail line was a proprietary system. While it was difficult or impossible to restrict use of roads to paying customers, railroad companies could charge the public for the use of the lines (either as passengers or as shippers of cargo) and use a portion of the funds generated to build more lines. Though improved roads were being built in England and the United States long before rail was invented, railroads quickly took over land transportation, and by 1869, a railroad line spanned North America. The settlement of the continent's interior that followed the railroad transformed the prairies not only by ending the buffalo migrations and turning under the native grasses, but also by introducing large numbers of invasive species across the entire region. Many of today's most damaging weeds— leafy spurge, St. John's wort, cheatgrass, and dozens of others—were introduced to the continent in the early 1800s, but became serious problems only toward the end of the century after the railroads had opened the continent to large-scale settlement. No longer did invasive species have to start their continental spread from peripheral coastal locations or along major river corridors. With the spread of the railroad, they were introduced throughout the continent's interior as well.

Because the tremendous range and cargo capacity of trains allowed them to distribute a single load of cargo to many locales over hundreds of miles of track, the development of a railroad system also made the containment of new pest outbreaks much more difficult and unlikely. Dutch elm disease, a beetle-borne fungus that kills American elms, was introduced to the United States around 1930 on several loads of unpeeled raw logs from Europe. At least three nearly simultaneous introductions occurred along the railroad route of transport that spread the disease across the heart of the continent. Within forty-five years, three-quarters of the elms in the northeastern United States had died.

The overland spread of nonnative species accelerated tremendously in the nineteenth century with the construction of a continentwide rail network. But since rail lines were far apart and the land between them was still the domain of horses and feet, the railroads saturated the continent with nonnative species only on a broad scale. The penetration of every valley with roads, however, has allowed for the distribution of nonnative species in detail.

The development of extensive road networks in North America followed the expansion of the railroad system by a half century or so. It was not until the late 1800s that road construction outside of built-up areas was seen as an important enough public good to warrant major public expenditure. By the turn of the century, though, the development of a rudimentary network of roads in the more settled portions of the United States had provided the impetus for the elaboration of a personal motorized transportation device, and by 1904, the Olds Motor Works was producing about 5,000 "automobiles" per year. The idea caught on, and by 1920, Olds's rival Henry Ford was building over 2 million Model T cars annually—a fortyfold increase in car production in just sixteen years.

Although trucks and cars have smaller hauling capacity than railcars, they have been produced in such large quantity that their aggregate capacity is enormous. Furthermore, even early cars and trucks had the power to go virtually anywhere even a crude road could be constructed. Seeding rivers and lakes with nonnative fish such as rainbow trout became a simple matter of loading them into trucks equipped with refrigeration and driving them into every valley that had a road. This is a job that could not have been done by railroads alone. Cars and trucks became America's endpoint distribution system, spreading nonnative species anywhere anyone wanted to introduce them.

Automobiles also serve as powerful vectors for the unintentional spread of insect pests and weed seeds. Many weeds are much more common along roads—even seldom-used four-wheel-drive tracks—than they are a few hundred yards removed from the road. Similarly, cars easily spread some kinds of pests long distances, starting new points of infestation. Gypsy moth egg masses, for example, are attached by female moths to many surfaces, including vehicles, lawn furniture, firewood, and campers—all of which may then be moved should the owner go on vacation to some distant site or move to another state. Similarly, aquatic invasive weeds that become tangled in boat props often get transported between lakes as boats are moved over the highway and then launched in another lake without first being cleaned.

The mechanization of transportation has made the introduction of new kinds of species much easier and much more likely as well. This expansion of the range of species types introduced is in part the product of increased shipping volume, for as the mechanization of transportation increased the capacity and range of cargo ships, there were simply more opportunities for somewhat less opportunistic species to move between continents as hitchhikers. However, the increased speed of international travel (both on steamships and, eventually, in planes) made possible the transport of species that simply could not have survived a long trip. Delicate species such as oysters, for example, could be shipped as larvae between continents. The European oyster *Crassostrea gigas* was introduced to Oregon and Washington in the 1920s to compensate for the decline due to overharvesting of the native oyster *Ostrea lurida*. These two factors—increased opportunity of transport and improved conditions in transit—acted synergistically to expand dramatically the pool of species that humans have moved from one continent to the next.

On the sea, the shift from sail to steam power just meant more of the same—more cargo in each boat, more boats in transit, more kinds of cargo transported. The first steam-powered ocean ships were built of wood, but their technological lifespan was short. By 1856, iron construction allowed ships to become enormously bigger, and from 1860 to 1880, iron steamships of up to 5,000 tons ruled the oceans. By 1880, though, shipbuilders were using steel instead of iron, and ships became even bigger and faster. In 1906, the steel-hulled ship *Mauretania* reached 32,000 tons, and by 1938, the *Queen Elizabeth* weighed in at over 83,000 tons. This tremendous increase in global cargo volume essentially opened a network of superhighways between ecosystems.

With mechanization, ocean travel changed from a dangerous occasional necessity to a normal part of business. Regular shipping allowed for greatly expanded trade, including the interchange of planting materials between various regions to improve or diversify agricultural operations. Apple cultivars, for example, could be shared between production areas in Europe, the Americas, Africa, Australia, and New Zealand such that each region became eager to gain production advantages by acquiring new plants from other regions. For this crop, like many others, the plant trade was associated with increased spread of pests. For example, San José scale—now a worldwide pest of apple—was once a minor pest restricted to the Russian Far East. In the last century, it has spread to apple-growing regions around the world, from New Zealand to North America, Europe, and South Africa, on apple nursery stock.

Increased travel also gradually changed people's perceptions of the role of nonnative species in their lives. The travels of wealthy Europeans and Americans often brought them in contact with new plant or animal species that they admired or felt had commercial potential, and plant importation became a major business. Europeans were particularly enamored of Chinese and Japanese plants, and in the nineteenth century a flood of Asian species poured into Europe and North America. Trees and flowering shrubs were especially likely to be imported. Some of the plants being moved around the world quickly proved to be invasive. *Lantana camara*, a tropical shrub from the Americas with beautiful flowers, was moved from the Americas to Hawai'i, Australia, New Zealand, southeast Asia, India, South Africa, eastern North America, and Europe between 1807 and 1924. In warmer climates, this shrub escaped to pastures and became a major pest of grasslands and woodlands.

In the second half of the nineteenth century, investors began to think increasingly in global terms, choosing places to start agricultural operations and the plants to cultivate there from a growing world inventory of available species and climates. For example, forestry operations in parts of the world with few or no native conifers (principally South America, Africa, Australia, New Zealand) began to screen the world pool of conifers to identify optimal species for local conditions and needs. Eventually, a group of about twenty species of pines, various spruces, firs, larches, and other trees came to form the standard set of timber species that many countries used to start forestry plantations, and around the world, mostly nonnative trees became the basis for the industry.

In short, the global shipping network made it far easier to search the world for the "best" species instead of using what nature provided at home. Those who had the resources to bring the world to their doorstep (or their garden) were relatively few in the late 1800s and early 1900s, but for the wealthy and the adventurous, seeking out the new and the unusual and bringing it home became a way of life.

While ships provide a link between ecosystems, another component of the developing world transportation system—canals—effectively merges two ecosystems into one. The first canals were built as early as the seventh century B.C., when Pharaoh Necho connected the Nile to the Red Sea by way of the Bitter Lakes. The Magic Canal in China connected two river systems over a watershed divide in A.D. 219, and the Romans built a canal between the Meuse and the Rhine in A.D. 619. All of these canals undoubtedly caused invasions between the ecosystems they connected, but it wasn't until the nineteenth century that canals began to be constructed on a grand enough scale that their effects became globally significant.

The Suez Canal was one of the earliest large canals built in modern times. Constructed by the French and completed in 1869, it provided sea-level passage between the Red and Mediterranean Seas, two formerly separate ecosystems with distinct biotas. Since the flow is primarily from south to north, a large number of Red Sea species have invaded the Mediterranean over the last 130 years. When the Panama Canal was completed in 1914, it created a transportation link between the Pacific and Atlantic Oceans. Since water flows outward in both directions from a higher-than-sea-level reservoir at the middle of the canal, there is no direct species movement between the two oceans. However, organisms can still be transported through the canal in ballast water or as growths of fouling organisms stuck to a ship's hull.

Even a canal that rises above sea level can increase the likelihood of certain kinds of direct invasions. In 1932, Canada completed the Welland Ship Canal, creating a link between the upper Great Lakes and the sea that formerly was blocked by Niagara Falls. Within twenty-five years, a parasitic fish called the sea lamprey invaded the lakes. Lampreys parasitize many kinds of freshwater fish and have decimated the Great Lakes lake trout population and the multimillion-dollar fishery it once supported. In the 1950s, the Canadian and American governments cooperated to build and upgrade canal and lock facilities in the St. Lawrence River to allow large oceangoing ships to reach the upper Great Lakes, creating the St. Lawrence Seaway. In the 1980s, ballast water in ships using this passage carried zebra mussel larvae into these lakes, precipitating a disastrous invasion.

Once a world ocean shipping network was established and a network of canals built, the only remaining factor restricting transportation of some species was the time the trip might take. Increasing speed of transit has been the special contribution of planes. The airplane was invented in 1903. By 1909, planes had crossed the English Channel, and a decade later a plane had flown across the Atlantic. The importance of speed is that stowaway species are more likely to live through the trip. A brown tree snake curled around the landing gear of a plane taking off from the island of Guam may still be alive hours later when the plant touches down in Hawai'i. The saving grace of air travel is that except for small private planes, the airplane transportation system is centrally controlled by a few large companies. This makes it relatively easier to adopt and enforce standards of what can be moved from one ecosystem to another.

Further increases in cargo capacity and shipping speed over the last half century have not only increased the likelihood that a species would be moved from one ecosystem to the next, they have also lowered the cost of international travel and shipping generally. With lower cost has come increased demand, and goods and people are now moving around the world at an ever faster rate. International trade has changed from a minor component of most countries' economic activity to a primary sector of the economy. This is the age of globalization.

## Homogenizing the Planet in the Age of Globalization

The world transportation net is now so comprehensive and so efficient that distance is largely irrelevant in the international movement of goods. In the global economy, it is no longer the method of transportation that sets the pace for biological invasions, but rather the choices made about what to move. Some countries have placed restrictions on the kinds of species that can be brought inside their borders (see Chapter 7). However, such regulations still have many gaps and at best have only slowed the rate of increase of introduction of new species.

Economic globalization directly affects the interecosystem movement of species in three basic ways. First—and probably most obvious—is the rapid rise in the overall volume of international trade. Second is an ongoing increase in the speed at which cargo and passengers are routinely moved from one part of the world to another. And third is the advent of new transportation technologies—such as containerized transport—that have the effect of enhancing pest survival, making pest detection more difficult, or both. The current state of these three factors represents more a change in degree than a change in kind, for all three have been more or less continuously on the increase since the beginning of human history. What makes the present situation unique is the unprecedented rate of acceleration in the volume, speed, and ease of international transportation—and the widening gap between the scope of the problem and the adequacy of the preventive measures put in place to date.

A graph of the change in volume of international trade since the 1950s looks just like a world population growth curve for the same period of time—it goes up almost exponentially. While the global gross domestic output is only five and a half times as big in 1997 as it was in 1950, the total volume of world trade increased sixteenfold in the same period. The curve is steepest in recent years, indicating that the gap between growth of global economic output and growth of trade is likely to continue to grow. One of the major drivers of the postwar growth of trade has been the ongoing dismantling of tariffs and other barriers to trade. The ecological consequence has been a vast increase in opportunities for species movement of all kinds as the volume of both potential pests and potential vectors rises.

International air cargo is a relatively new phenomenon, but this high-speed mode of shipment is growing exponentially as well. In 1989, there were only three airports in the world that handled more than 1 million tons of air cargo. Just seven years later, the number of major air cargo hubs had increased to thirteen. The risk with such high-speed transportation of freight is that fragile species that would not have survived a longer ocean voyage—some insect pests, for example, or microbes of various kinds—are now much more likely to participate in the global mixing of ecosystems.

Ports were for many years the point of origin for most unintentional alien species introductions, and as a consequence, both numbers and dominance of invaders tended to decrease with distance from harbors, airports, and other major hubs in the international transportation system. The invention and almost universal adoption of containerized transport, however, has in the last few decades begun to spread accidental introductions across the entire landscape. Instead of unpacking cargo (and therefore releasing most stowaway pests) at the dock, containers are now loaded on trucks or railcars and shipped to their various destinations unopened. This shift has dramatically increased the likelihood that initial outbreaks of new pests will occur simultaneously in many locations across the country, making eradication nearly impossible.

As alarming as these direct effects of globalization might be, the indirect impacts of recent changes in the world trade system are even more profound. Chief among these impacts are the increasing numbers of species that are being spread around the globe as a consequence of the dramatic expansion in recent years of the plantation timber, aquaculture, and large-scale agricultural industries in developing countries. The difference between these impacts and those mentioned above is that many of the commodities that these industries export do not in themselves present an unusually large risk of causing an invasion. Rather, the demand for these commodities—almost all of which are exported to industrialized nations—provides a strong economic incentive for deliberate introductions in developing countries that participate in the global trade system.

For example, shrimp farming has become a major industry in developing countries because declining wild fisheries and growing demand in industrialized countries ensure a steady market for frozen shrimp. While a shipment of frozen seafood is highly unlikely to harbor pests, the shipment of live shrimp breeding stock from one farm to another is spreading an epidemic of aquatic wildlife diseases across the globe. Christopher Bright, a research associate at the Worldwatch Institute who studies biological invasions, describes shrimp farming as "a form of 'managed invasion.'" Given the important role of shrimp and other small crustaceans in marine food webs, these disease invasions have the potential to cause the collapse of entire ecosystems.

Plantation forestry and large-scale export agriculture have had similarly disastrous consequences, though for different reasons. Tree plantations in developing countries generally use nonnative species, a substantial number of which have proven quite invasive. Almost all of the recent growth in plantation forestry in the developing world has been fueled by demand from industrialized nations. A similar phenomenon is at work in export agriculture—particularly fish farming—where major new operations are introducing a wide variety of species to aquatic ecosystems throughout the developing world. This trend is also largely a product of the recent opening of agricultural markets and of the subsequent demand in industrialized countries for cheap foreign-raised fish such as tilapia and Chilean farm-raised salmon.

The consequences of globalization extend beyond the piecemeal transportation of species as commodities or stowaways. The global economic system and the transportation system that makes it run are driving changes in local economies that have overwhelmingly tended to favor the mixing of ecosystems everywhere. Western civilization has developed the means to move anything anywhere and the incentives to do it. The only constraint left is good judgment. The challenge is to develop this judgment and the mechanisms to enforce it in time to stem the homogenization of the world.

PART TWO

# Saving Nature from Nature

## Strategies for Preventing and Controlling Biological Invasions

# CHAPTER 4

# Refuge for the Mussels

*Biotic Integrity and Zebra Mussel Invasion
in the Ohio River Basin*

Our boat seemed to be the only thing moving on this entire stretch of the
Ohio. The bright August sunshine made the river a mirror, quiet and impas-
sive all the way to the Kentucky shore a half mile away. As we pulled away
from the boat launch, the surface of the water split cleanly across the prow,
pushing smooth, even ripples out in two long arcs behind the boat. There
were no rocks, no riffles, no eddies in the current—almost nothing to indicate flow. Though the map said river, my eyes said lake. There is a dam every
50 miles on average on the Ohio, and in most places, nothing but a map can
tell you which way is downstream.

But Patricia Morrison was at the helm, and she knew where she was
going. We were headed a mile or so westward (downriver, she said) to a particular stretch where she knew there was a large gravel bar 10 or 15 feet
below the surface. As a biologist for the Ohio River Islands National
Wildlife Refuge, Patricia was responsible for keeping track of these gravel
bars hidden beneath the water. These were the only places where the river's
remaining native freshwater pearly mussels survived, and it was Patricia's job
to monitor their numbers and their health.

We pulled up over the place where the mussel bed had been last year and
Patricia dropped the anchor. There were two other people with us suited
up in scuba gear, ready to get to work. But first Patricia had to do some sampling to make sure the gravel bar was still where she had left it—that it had-
n't shifted in a storm or been buried by sediments. She dropped a mussel-
catching contraption called a brail over the side and dragged it along the

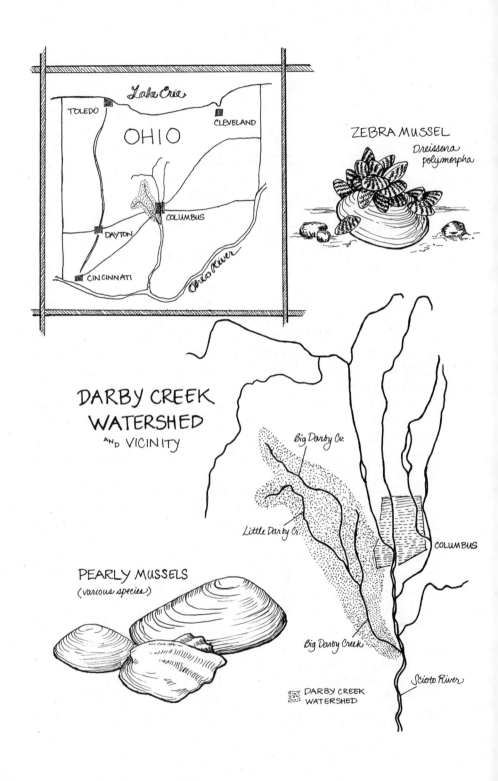

OHIO

Lake Erie

TOLEDO

CLEVELAND

DAYTON

COLUMBUS

CINCINNATI

Ohio River

ZEBRA MUSSEL

*Dreissena polymorpha*

DARBY CREEK
WATERSHED
AND VICINITY

Big Darby Cr.

Little Darby Cr.

COLUMBUS

PEARLY MUSSELS
(various species)

Big Darby Creek

Scioto River

DARBY CREEK
WATERSHED

bottom. It came up with a couple of mussels attached. "This is it!" Patricia exclaimed, and her two divers dropped overboard.

They came up five minutes later with the first of many baskets of gravel from the bed below. Using a protocol Patricia had developed when they began monitoring a few years before, the divers systematically dug up and sent to the surface a series of quarter-meter-square samples of the top 5 inches of the gravel bar. We dumped the buckets out on a large metal tray and began scanning for mussels. Most of what came up was gravel of various sizes, from pea-sized pebbles to cobbles the size of grapefruit. And then a pearly mussel. "A fawn's foot," Patricia said with a grin. There are hundreds of species of these pearly mussels in eastern North America—more than in any other place in the world—and most of them have names equally imaginative. The early settlers who named them were inspired by the diversity of shapes and colors in the mussels they found, from inch-long purple beauties with paper-thin shells to pearly white giants whose dinner plate–sized shells were almost a quarter-inch thick. No wonder Patricia smiled every time we found a mussel among the rocks.

But these pearly mussels—known scientifically as unionids—were not the only kind of mussel that came onboard with the buckets full of gravel. For every native the divers brought to the surface, they sent up hundreds of

Biologist Patricia Morrison and a colleague separate native unionid mussels from invasive zebra mussels in a sample taken from the bottom of the Ohio River. (Jason Van Driesche)

Six of the many species of pearly mussels native to North America. (Kevin S. Cummings, Illinois Natural History Survey)

tiny alien zebra mussels as well. These brownish-drab creatures with distinctive stripes on their shells were accidentally introduced to the Great Lakes in the late 1980s in ballast water of ships from Europe and have been spreading through the major waterways of the eastern United States at a tremendous pace ever since. During the previous year's sampling, Patricia's divers had found about 1,000 zebra mussels per square meter here—up from fewer than 20 per square meter only two years before that. "This seems to be the leading edge of zebra mussel reproduction," Patricia said. "We saw a thirty-fold increase on this site in just one year." But 1,000 per square meter is only the beginning. A few hundred miles downriver—where the invasion took hold a couple years earlier—there are now as many as 30,000 zebra mussels per square meter of riverbed.

These billions of invaders in the Ohio River have proven to be anything but good neighbors. As adults, zebra mussels are sedentary, attaching themselves to any hard substrate with dozens of thin, sinewy cords called byssal threads. For some reason, they seem to prefer native mussels over rocks as substrate. Most of the native mussels we pulled up from the bottom had at least a half-dozen zebra mussels attached to them, and some had many more. "This is double trouble for the natives," Patricia explained. "Not only are the zebra mussels competing with them for food, but they're weighing the natives down as well. A unionid can accumulate such an incredible load of zebra mussels that it loses all maneuverability." She often comes across native mussels that are carrying around more than their own weight in hitchhikers.

The weighted-down natives generally fail to reproduce, for the sperm that the males release into the water column are likely to be filtered out before they ever reach a female unionid if she is covered with a ball of zebra mussels. Zebra mussel–encrusted natives also use up so much energy dragging around their load of hitchhikers that they are often unable to store enough energy to take them through the winter. As a consequence, Patricia has seen the same pattern repeat itself in one site after another: As zebra mussel populations climb, native mussel reproduction ceases almost immediately, and within a year or two mortality goes through the roof. At the most heavily infested sites, she has seen up to 70 percent of the native mussels die in a single year.

The spread of zebra mussels is the single largest threat to the survival of native mussel species in the Ohio River and across much of North America. Even before the introduction of zebra mussels, unionids were already one of the most threatened groups of species on the continent, with 60 percent of the 297 recognized species considered threatened or endangered and 12 percent believed extinct. But the zebra mussel invasion has accelerated the slide

This native unionid mussel is carrying a load of dozens, perhaps hundreds, of zebra mussels. (Courtesy of Illinois Natural History Survey)

of native unionids toward extinction dramatically, and one group of researchers has predicted that their invasion will combine with ongoing habitat degradation to push about 60 more species of native mussels—another 20 percent of the total native mussel fauna—into extinction by the early part of the twenty-first century. Most of these extinctions will probably occur in lakes or in big rivers like the Ohio.

Emergency action clearly is needed, but thus far there is little that conservationists can do except watch and worry. Eradicating zebra mussels is obviously impossible, and effective control is unlikely on anything more than a very limited scale. Chemical control holds little promise, for any chemicals that kill zebra mussels would also kill unionids. And while research into biological control is under way, almost no examples of successful biocontrol of aquatic invertebrates exist. There is no quick fix for the zebra mussel invasion—no easy cure for past mistakes.

The problem is that zebra mussels are perfectly adapted to the kind of aquatic habitat that people create: deep water that moves slowly or not at all. Zebra mussels love dams and dredging almost as much as people do, and the link between rivers modified by humans and the spread of zebra mussels is indisputable. The zebra mussel invasion is yet another indication that nonnative species problems are often just as much a symptom as a cause of ecosystem degradation. The only way to mitigate their impacts is to stop turning rivers into lakes—and, where possible, to restore the original character of rivers that have been dammed and channelized beyond recognition. "If we let this river behave more like a river," Patricia said, as we searched for natives among the invaders and the rocks, "zebra mussels wouldn't be a problem here."

## The Perfect Invader

Zebra mussels have adapted well to a world remade by humans. From their native home in the rivers that drain into the Black, Caspian, and Aral Seas, they have spread out across the globe over the last couple of centuries as hitchhikers on and in ships of all kinds. Their range expansion began with the growth of commerce between eastern and central Europe in the sixteenth century, and by 1790, zebra mussels were reported in the River Danube in Hungary. Over the course of the next century, they made their way westward through the inland waterways of the continent, reaching the Atlantic seaboard in the late 1800s. By the turn of the century, nearly every navigable river and lake in Europe south of Scandanavia was saturated with zebra mussels.

This early range expansion was largely realized by adult zebra mussels attached to the hulls of boats that plied the canals built between Europe's rivers over the course of the nineteenth century. Shipbuilding technology prevented the spread of zebra mussels beyond Europe until the turn of the century. Because zebra mussels are a freshwater species, they couldn't survive a transoceanic voyage in salt water. It wasn't until the early 1900s that steel ships began using water as ballast—thus carrying river and lake water internally from one continent to another. Since zebra mussel larvae are free-floating, they are readily taken up if they are present when a ship takes on freshwater ballast.

But as the Industrial Revolution spread across Europe, increasing pollution in inland waters caused zebra mussel populations to fall dramatically. Though zebra mussels are tolerant of much lower water quality than are many other aquatic invertebrates, they do not thrive in highly polluted waters. As a consequence, although shipping between European and North American ports has provided a vector for the movement of zebra mussels between the two continents since the turn of the century, for many years there were too few mussel larvae in Europe's polluted rivers and harbors for ballast water to serve as a source for invasion. Since many of North America's waters—particularly its major harbors—were equally polluted, the few mussels that made the voyage were unlikely to survive once they arrived.

Ironically, the environmental movement is indirectly responsible for the establishment of zebra mussels in North America. By the early 1980s, water quality had improved significantly in the inland waters of both Europe and North America, and both the source and the receiving environment were far more hospitable to zebra mussels than they had been fifty years earlier. The first zebra mussel report in North America was from Lake St. Clair, a small body of water that connects Lake Erie with Lake Huron. The infestation was discovered in 1989, but given the size of the mussels found, it was estimated that the first introduction must have taken place sometime around 1986. Zebra mussels spread rapidly once they became established in North America, either as adults attached to the hulls of ships or as larvae in water currents. By the end of 1990, they had infested the entire Great Lakes basin.

For the first few years, researchers and government officials tracked new invasions closely, but it quickly became apparent that nothing was going to stop their spread. By 1992, zebra mussels had invaded the Mississippi River drainage through canals that connect that system with the Great Lakes and began moving up the Mississippi's tributaries with barge traffic. Given how quickly they have continued to spread, it appears likely that zebra mussels will have invaded virtually all suitable habitat from northern Florida to central Canada and from coast to coast by 2010.

The rapid growth and spread of the zebra mussel infestation stems in part from their enormous reproductive capacity. A single adult mussel can produce as many as one million larvae per year, and each one of those larvae reaches reproductive maturity in less than twelve months. More important, the reproductive cycle of zebra mussels is much simpler and more flexible than that of native mussels. Each species of unionid depends on one or several particular species of native fish that serve as hosts for its larvae, creating an ecological relationship between fish and mussel that limits both the geographic range and the reproductive capacity of every native mussel species. Zebra mussels suffer no such complications; their larvae are free-floating, allowing the species to reproduce in any suitable environment. This means that zebra mussel reproduction is constrained by much broader parameters— principally, food resources and water temperature—than is native mussel reproduction, allowing zebra mussels to invade and multiply rapidly in a wide range of habitats independent of the presence or absence of a larval host.

In one place after another, zebra mussel densities have gone from 10 to 10,000 mussels per square meter in two years or less, and in particularly favorable environments, densities have gone as high as 750,000 mussels per square meter. This is nothing short of a biological explosion, and native mussels have been caught in the blast from the start.

## The Engineering of a River

The day's sampling was done. Patricia pulled up the anchor and dumped the last of the gravel and mussels back overboard. As we pulled away from the gravel bar and headed back upstream, I glanced at the depth finder on the console. Fifteen, twenty, thirty feet—the river got progressively deeper as we moved into the channel. Patricia noticed me watching the river bottom fall away. "You may find this hard to believe," she said, "but the average depth of the upper Ohio River prior to impoundment was about a foot."

When the first European settlers began trickling into the Ohio River valley in the mid-eighteenth century, they found a river so shallow that some would walk across it every day to farm or hunt on the other side. The water wound through gravel bars, over flood-downed snags, and through riffles and occasional deep pools, dropping an average of a little less than half a foot per mile for the 981 miles from its origin at Pittsburgh to the confluence with the Mississippi at Cairo, Illinois. Except for a few months of high water in the spring and fall, navigation on the Ohio in those days was a dangerous and uncertain business. According to one account from 1819, "the captain of a boat drawing only fourteen inches reported it took him thirty-five days to navigate from Pittsburgh to Cincinnati because he grounded fifty times on shoals where the river was ten inches deep." Nine months out of the year, the Ohio was more an oversized stream than a navigable river—a natural state that before long would prove incompatible with human demands on what was to become one of the new nation's most important waterways.

The Ohio's shallow riffles were ideal habitat for many species of freshwater mussels. At the time of European settlement, there were about 80 species of mussels in the river's main stem and 127 in the watershed as a whole— the most diverse freshwater mussel fauna of any river basin in the world. (Compare this to Europe, where there are only about 60 species native to the entire continent.) Shallows in the Ohio probably harbored hundreds of mussels per square meter, and anyone who walked across the river walked as much on the backs of mussels as on gravel. Although a sizeable pearl button industry made use of the shells in the nineteenth and early twentieth centuries, the commercial value of the mussels themselves paled beside the value of the river as a "common commercial highway." As a consequence, anything that could be done to improve the river's usefulness as a transportation corridor generally received strong support, regardless of the consequences for the mussels.

In 1824, the Army Corps of Engineers began channel clearance and "improvement" in the Ohio, removing snags, blasting out boulders, and installing "wing dams" to deepen the channel. These early structures were not actually dams, but rather angled barriers protruding from both sides of the main channel that acted as a funnel, deepening the channel by concentrating the river's flow. This work was the first large-scale modification of the basic form of the riverbed. Over the next fifty years, wing dams were used in combination with dredges to create a minimum low-water channel depth of 3 feet in the lower reaches of the Ohio and 2.5 feet in the upper river.

For the most part, these early efforts simply made the original Ohio a little safer for navigation without greatly altering the basic character of the river, and mussels remained abundant well into the second half of the 1800s. As the century drew to a close, though, several factors came together to begin a process of change that in the twentieth century had obliterated the old river and replaced it with a new one, decimating the native mussel fauna in the process.

The Ohio's first slackwater dam (one that holds back the current to create a pool) was built just below Pittsburgh in 1885, providing a harbor for the city and easy passage for barges through its locks. The project proved so popular that, in 1910, Congress approved funding for the "canalization" of the river with an additional fifty-three dams along its entire course. Over the next two decades, the Army Corps of Engineers built fifty of these dam-and-lock structures, opening a 9-foot-deep slackwater channel from Pittsburgh to the Mississippi. These "wicket dams" were built like the tailgate of a pickup truck, with a series of large wooden panels that were attached by hinges to a concrete foundation on the riverbed. When the river was low, the panels were propped upright to create slackwater pools for navigation; when high water came, the panels were lowered to the riverbed and the river flowed free. Though some shallow side channels still existed on the Ohio in which native mussels could survive even after the completion of the canalization project, the creation of a 9-foot-deep channel destroyed most of the river's riffle habitat. Combined with the sedimentation that followed land clearing for agriculture, this channelization project initiated a precipitous decline among many of the Ohio River's native mussels.

Deforestation and land clearing also contributed to a series of devastating floods and droughts in the Ohio in the early twentieth century. Public demand for government-sponsored flood control projects grew throughout the 1920s, and following the establishment of the Tennessee Valley Authority in 1933, the federal government began constructing high dams in the basin. Though their initial purpose was flood control, it soon became apparent that dams were also useful for maintaining channel depth in the Ohio and its navigable tributaries during periods of low water. In the 1930s, the Corps of Engineers designed and was authorized funding for a total of seventy-eight dams on the various tributaries that fed the Ohio, almost all of which were eventually built. The overall effect of these dams on the river's ecology was to make its aquatic habitats much more uniform in both time and space. According to a recent history of the engineering of the Ohio, "the river's floods are not so high, nor its low flows so puny. . . . The resulting flow changes on the Ohio inevitably affect its riverine environment, over time establishing an ecology differing from that existing before the flood control program began."

By the 1950s, the small size of the original wicket dam locks and the large number of lock passages required along the course of the river had become a constraint on an ever-expanding river transportation system, and the Corps began planning for a smaller set of larger fixed dams that would replace the antiquated wicket dams. By 1999, fourteen of these higher dams had been built and the last was under construction, replacing the low dams and creating a series of wider, deeper pools that submerged much of the remaining shallow-water habitat. This undertaking was known within the Corps as the Ohio River Modernization Project, for it upgraded the river "from a local into an interstate highway for commerce." When the final dam is completed in 2005, the Corps will have finished its 175-year-long conversion of a highly diverse river environment of riffles and pools and floods and droughts into a series of reliably uniform lakes. The benefit is measured in cheap transportation—and the cost in extinctions of pearly mussels and the death of a river.

Though from a human perspective the conversion of the river has happened quite gradually, the change was far too abrupt for most of the river's pearly mussels. Even before zebra mussels arrived, the combined effects of habitat loss, pollution, and siltation had put more than half of the continent's unionid mussels on the threatened and endangered species lists and had caused at least thirty-five of them to go extinct. Losses were greatest in

A series of dams built by the Army Corps of Engineers has transformed the entire length of the Ohio River from a highly diverse ecosystem of riffles and pools into a staircase of uniform lakes.

large rivers like the Ohio, where dams had altered not just portions of the river but its entire run. Most of the early mussel extinctions were of species that lived exclusively in shallow riffle habitat in big rivers—the first kind of river habitat on the continent to be subjected to intensive, comprehensive and continual human modification.

In a series of studies of ecosystems under stress, ecologist David Rapport has found that this process of ecosystem modification is often characterized by continual simplification of the natural system in question. In the Great Lakes ecosystem, for example, human modification tended first to destroy bays, shoals, wetlands, and other highly productive areas, transforming complex, well-integrated systems into simple, disorganized ones. In many cases, nonnative species invasions have followed and exacerbated degradation, and while it is rarely possible to establish definitively that habitat degradation allowed an invasion that otherwise would not have happened, there is a strong relationship between habitat degradation and nonnative species invasions. From an ecological point of view, people are simply the most catastrophically damaging invader of all time, and it makes sense that the conditions created by one opportunistic, highly adaptable species—Homo sapiens—would facilitate invasion by another.

This transformation of the Ohio from a varied environment of riffles and pools to a nearly uniform environment of still, deep water predisposed the river to massive invasion by zebra mussels. Millions of research dollars invested in a search for zebra mussel control methods have only shown that fighting the invasion directly is useless. According to David Rapport, this is typical of ecosystems under stress. He believes that "too much effort has been expended in trying to regulate the system, instead of the human activities that stress and transform the system." The best way to save the native mussels of the Ohio from obliteration by zebra mussels is to remove the human-induced stresses—deep water, slow current, sedimentation, and pollution—that have made the Ohio so inhospitable to unionid mussels and other native species and so favorable to zebra mussels and other invaders.

The Fish and Wildlife Service has begun to work with the Corps of Engineers to investigate the possibility of altering the release schedule from the headwaters dams periodically so as to simulate flood conditions and sweep zebra mussel larvae out of the system. This approach would begin to address the cause rather than the symptoms of the zebra mussel invasion. Restoring water clarity and seasonal variations in flow is an essential first step in the preservation of the Ohio's remaining pearly mussels. Otherwise, one new invader after another is likely to take advantage of the altered state of the Ohio, until nothing is left but barges and aliens.

Restoring the Ohio is worth more effort, but too much damage has been done to this river to hope in the short term for anything more than a slowing of losses. This kind of "dams versus mussels" dilemma is one that, from a conservation standpoint, looks rather hopeless. The best immediate investment of conservation money and energy is in those places that have yet to be significantly altered. All else being equal, saving these kinds of places is probably the best insurance against invasion.

## A River That Doesn't (Yet) Need Fixing

Read these names aloud:

Paper pond shell. Giant floater. Cylindrical paper shell. Squaw foot. Elk toe. Slipper shell. Simpson's mussel. White heel-splitter. Hackle-back. Creek heel-splitter. Washboard. Buckhorn. Maple-leaf. Rabbit's foot. Pimple-back. Blue-point. Wabash big-toe. Purple pimple-back. Northern club-shell. Ohio pig-toe. Lady finger. Horn-shell. Kidney-shell. Hickory-nut. Deer-toe. Fawn's foot. Fragile paper shell. Pink heel-splitter. Fragile heel-splitter. Liliput shell. Black sand-shell. Bean-shell. Rainbow-shell. Fat mucket. Pocketbook. Small mucket. Snuffbox. Northern riffle-shell.

These are the thirty-eight species of native mussels that presently live in the Darby Creek watershed in central Ohio. The creeks and streams that make up this system are as beautiful and as surprising as the names of the mussels that inhabit them. Though the Darby watershed is overwhelmingly dominated by farmland, it is one of the cleanest, most diverse river systems in the Midwest. Nearly all of the native animals and plants that were found here at the start of European settlement are still present in the Darby, for the river and its tributaries run shallow and remarkably clear over most of the 90-odd miles from the headwaters to the mouth. In this part of the country, this is about as far as you can get from the dams and the slackwater of the Ohio River.

The Darby is by no means typical. Almost two-thirds of the small- to medium-sized streams in Ohio are moderately to extremely degraded, with an Index of Biotic Integrity (IBI) of under 40. (The IBI is a measure of the overall health of a stream, as indicated by the composition and abundance of its fish populations. The more species and individuals there are of the most environmentally sensitive fish, the higher the rating.) The Darby watershed is near the top of the category called "exceptional." In a state ranked third worst in the nation for degraded waterways, it is surprising this category exists at all in Ohio.

How the Darby escaped degradation will always be something of a mys-

Darby Creek's clean, shallow water is some of the best native mussel habitat left in the state of Ohio. (Courtesy Steven Flint, The Nature Conservancy)

tery. According to Jack McDowell, a retired high school biology teacher who spent many years on the Darby with his students, fortuitous topography has played a significant role in keeping the river system intact. Many of the tributaries in the western portion of the watershed are spring-fed, which provides a continuous source of water whose quality is relatively unaffected by land use. In addition, many of the Darby's tributaries—particularly those that drain into the Little Darby—are lined with low bluffs that are set close enough to the creeks that farming the land between the bluffs and the water is not worth the bother. As a consequence, a riparian corridor has been maintained along most of the creeks despite intensive cultivation of surrounding lands.

Land use has in two respects played a part as well. First, the watershed was historically dominated by small, diversified farms whose owners kept the steeper, more erodible lands that lay just above the riparian corridor in pasture. It was only in the 1960s that farmers began to plow these pastures under—a trend that just thirty years later is being corrected through the Conservation Reserve Program. Second, settlement patterns in Ohio are such that the Darby Creek watershed is one of the few drainages of any size that has remained almost entirely rural. Though the rapidly growing city of Columbus lies immediately to the east of the Darby, there are no large municipalities entirely inside the watershed. This is one of the few creeks in Ohio that—until recently—no one had bothered to change.

But the pressures on the watershed are tremendous. According to the U.S. Census, the Columbus metro area is projected to grow by at least 250,000 people before the year 2015. Both Columbus and the nearby city of Hilliard have already annexed land within the drainage of the Darby, and land values in many parts of the watershed are soaring. It is uncertain how much longer the area will remain mostly rural.

Steven Flint is well aware of just how important this watershed is. As River Steward for The Nature Conservancy's Darby Project, it is his job to serve as the project's eyes and ears over the 560 square miles of land that feed the Big Darby, the Little Darby, the Spring Fork, Hellbranch Run, and all the other streams in the watershed. "I've walked the main stem and most of the key tributaries, and I know every inch of them," he said. "I can get out there in a heartbeat if there is an incident anywhere."

Steven and I were out on a day in mid-June investigating the consequences of what was being described as a 200-year flood. A week or so earlier, 11 inches of rain had fallen in a little over forty-eight hours, and the creek burst out over its banks onto the floodplain. Now that the water had receded to near-normal levels, Steven was impatient to see what was going on down at the creek. We stopped at a bridge that crossed the Darby somewhere in its middle reaches and headed down to a gravel bar just upstream. The water was still a little cloudy, but it had cleared up considerably since the height of the flood. Steven squatted down at the edge of the flow and poked around in the gravel beneath the water's surface. After a few seconds of rummaging, he pulled out a pearly mussel the size of his fist. "Looks like they came through just fine," he said with a grin.

Not that this is any surprise. The fortuitous combination of active stewardship and benign neglect that has allowed this watershed to retain an unusual measure of health and resilience means that the Darby will be running clear again within a few days. Not only have the Darby and its tributaries been largely insulated from the effects of changes in the land since settlement; the creek itself is remarkably free of modification as well. (There are only two very small dams in the entire watershed.) The ecological web that connects mussels to fish to clean water to everything else is largely intact. This is an ecosystem that has never even approached the level of degradation that the Ohio River passed many years ago.

What's more, the Darby is the kind of swift, shallow stream that zebra mussels probably will never overrun. According to Tom Watters, a zoologist specializing in mussels at the Ohio State University in Columbus, it is unlikely that zebra mussels have even been introduced to the Darby because the undeveloped state of the river discourages motorboat use (which is the principal vector for interwatershed movement). But even if they are eventually introduced—and this is the most important point—they will probably fail to establish there, or would at most establish only at low levels in the occasional deep, still pools that occur naturally along the Darby. And every time the creek flooded, most of the zebra mussel larvae in the water column would get washed right out of the watershed and down to the Ohio. The Darby simply isn't hospitable to zebra mussels, for these invaders are generally much more at home in the kind of river habitat that humans tend to create—deep, slow water that rarely if ever floods—than in the kind of habitat that exists in many of the rivers we leave alone.

The beauty and the ecological value of the Darby have earned it designation as both a State and a National Scenic River, and it is unlikely that it will be dammed or channelized at any point in the foreseeable future. Even so, there are serious threats to the river's integrity—threats more insidious than dams or other direct habitat modifications. Most significant among these are poor agricultural practices and suburban sprawl.

In the late 1980s, The Nature Conservancy initiated a campaign to address these watershedwide threats to the health of the system with the establishment of a fifty-member coalition of concerned agencies and organizations called the Darby Partners. According to Steven Flint, the Darby Partners is intended to be a human "web" of management and care that is as complex—and hopefully as strong—as the ecological web of the Darby ecosystem itself. The group has helped locate and clean up undocumented agricultural and septic outfalls that drain into the Darby. It has promoted the use of vegetated buffer strips, no-till planting, and other soil conservation measures on agricultural land. It has steered development away from the watershed wherever possible and pushed for strong mitigation measures wherever development was inevitable. Most important, it has encouraged a strong working relationship among the many different interests—farm organizations, environmental groups, city planners, and others—that have a stake in the health of the Darby watershed.

This diverse membership is what gives the Partners its strength, but the Conservancy is what holds it all together. And according to Tim Richardson, the urban development coordinator for the project, Steven Flint is what keeps the Conservancy in touch with what this project is all about—the creek itself. "I think that the most important thing that The Nature Conservancy has going for it is Steven Flint," Tim said. "That guy is out on the creek sometimes seven days a week. He knows what's going on out there. If anything is going to save this system, it's the constant monitoring that Steven does."

Steven and I soon left our mussel hunting on the main stem of the Darby and headed out to the Spring Fork, one of the highest-quality tributaries in the watershed. Steven took a while getting us there, though—there were a dozen places he wanted to show me along the way, and every corner we came around reminded him of something more that he had to point out. Most of the trip was a running commentary on agricultural practices in the fields we passed. "What you're seeing there is land in CRP—the Conservation Reserve Program," he said as we passed a field filled with tall grass and cone-flowers. Over the next rise was an eroded low fold in a field, and Steven shook his head. "These areas here—these might be good candidates for a grass swale," he said. In his mind, everything led down to the water.

A few minutes later, we crossed the Spring Fork on a small bridge and made an immediate right into the driveway of Dale Rapp's 1,200-acre farm. Mr. Rapp still grazes and plants the upper part of the property, but around 1995 he began fencing cattle out of the land along the creek. "He noticed the bank erosion and the obvious nutrient loading from letting a dozen head of cattle get into the stream for an afternoon, and just wanted to improve things," Steven explained. "That's good stewardship." Though the Spring Fork drainage has an even higher percentage of land in agriculture than the watershed as a whole—92 percent as compared to 80 percent or less—the creek still rebounded dramatically from the impacts of cattle. Of the 103 species of fishes native to the watershed, upwards of 60 are now found in the Spring Fork alone.

We walked down the pasture and through the gate to a graveled path that ran down to the creek. The bottomlands were chest-high in grass, and young trees grew thickly along the bank. Steven squatted down by the water's edge and pointed out into the flow of the water. "Look at the substrate," he said. "It's almost silt-free—more sandy glacial till than heavy clay sediments. And this is after that rain we had last week." The riffles and rocks meant that this stream was structurally intact. It was just what native fishes and unionids need—and just what zebra mussels are ill-equipped to handle. Mr. Rapp was taking care of the Spring Fork.

Many other farmers in the watershed are making a similar effort to pro-tect riparian areas and the stream itself from the impacts of agriculture. Their concern was inspired in part by a recognition that unless farmers took proactive steps to improve agricultural practices and reduce their impacts on the creek, the inevitability of regulation in an ecologically valuable water-shed would soon force change upon them. In 1991, a small group of farmers from throughout the watershed came together to discuss the implications of the Darby's ecological status for farming in the watershed and to design a response to the environmental community. The product of their meetings was Operation Future, an organization characterized by its founders as "farmers protecting Darby Creek and the bottom line."

One of Operation Future's first activities was a canoe trip on the Darby in which an environmentalist was paired with an Operation Future member in every canoe. They couldn't help but talk to each other, and the trip has since become an annual event. Dennis Hall, the executive director of Operation Future at the time, believes that the first trip triggered a "fundamental shift" in the relationship between farmers and environmentalists in the watershed. "The notion that biological diversity had persisted in this agricultural watershed while not in others was recognized as testimony to the farmers' stewardship," Mr. Hall explained. "A new sense of pride and ownership of the creek's biological diversity began to emerge in the farming community. Farmers began to see their role as making a great place even better." The farmers had gone into the trip expecting to be criticized for agriculture's contribution to water-quality problems and instead found an ally in keeping the watershed in good shape.

Since that first trip, nearly two hundred farmers have joined Operation Future, and together they farm about 30 percent of the most critical lands in the watershed (those that are adjacent to waterways). These and other farmers have worked with a wide variety of organizations to implement soil conservation practices such as buffer strips and minimum tillage, and over about five years they had reduced the amount of sediment entering streams from agricultural fields by about 35,000 tons per year. The organization launched a major membership drive in 2000 and hopes to bring in half of the farmers and three-quarters of the land in the watershed by the end of the year. Operation Future has proven as popular as it is effective, and participating in its conservation-related mission is now socially acceptable in the farming community. The group's accomplishments have shown that agriculture can be entirely compatible with the maintenance of the kind of high-quality streams that favor native mussels and resist invasion by zebra mussels and other nonnative species.

Development, on the other hand, is past a certain point essentially incompatible with watershed health. Land on the eastern edge of the Darby watershed is being converted to suburbs for the booming city of Columbus at a tremendous rate, and the impacts on the creek are obvious. The most damaging consequence of development is the increase in rapid, unfiltered runoff from impervious surfaces—streets, sidewalks, roofs—that prevent

Changes in surface runoff patterns following suburban development are a major threat to the integrity of the Darby ecosystem. (Courtesy Steven Flint, The Nature Conservancy)

rainfall from percolating slowly through the soil and instead send runoff straight into the nearest stream. According to Steven Flint, experience in other watersheds has shown that once impervious surfaces constitute 10 percent of the land cover in a watershed, stream health begins to falter. At 15 percent, serious degradation is widespread, and by 20 percent, Steven said, "You're in real trouble." Above 20 percent, all but the hardiest (often non-native) species are gone, and there is little hope for recovery. Though the Darby watershed as a whole is still less than 10 percent impervious surfaces, some sections are approaching that threshold. And the damage is beginning to show.

Lake Darby Estates is one of the oldest subdivisions in the watershed. It is also the closest to the creek. Immediately to the west of the development is the Conservancy's Fox Trap Preserve, a 211-acre parcel of former farmland along the main stem of the Darby. All of the storm drains for the 115-acre development are routed through a single large culvert that discharges into the preserve. "Let's take a look at just how much damage this stormwater outfall has caused," Steven said as we walked down from the subdivision into the preserve. "I've nicknamed this one 'son-of-a-ditch.'"

And a ditch it was. The mouth of the pipe jutted out over a ravine 15 feet deep and 60 feet wide that shot straight down though the woods toward

The infamous "son-of-a-ditch," where unmanaged stormwater from a 115-acre development has sent thousands of cubic yards of sediment into the Darby. (Courtesy Steven Flint, The Nature Conservancy)

the Darby. Steven pointed to the lip of the culvert. "When they put this subdivision in, that pipe was only a foot and a half above ground level," he explained. The entire volume of the ravine that gaped below us—thousands upon thousands of cubic yards of soil—has washed down into the Darby.

We clambered and slid down to the bottom of the ravine, and picked our way through an assortment of old tires and broken appliances and shreds of plastic bags. A few years ago, the Conservancy attempted to slow the passage of water by anchoring a series of tractor tires to the bed of the ravine with heavy-duty steel posts. The first major rainstorm blew them all out. "It's just one of those things we've learned to deal with," Steven said. "There's really nothing we can do about it at this point." The Darby's Index of Biotic Integrity drops sharply for several miles below the ditch, and the stormwater discharge from this subdivision is largely responsible.

The same thing is happening in the Darby's headwaters. Honda Motor Company's main North American plant covers thousands of acres in the northern reaches of the watershed, and the runoff from its enormous parking lots and from the roof of its plant drain into the Flat Branch. This small tributary of the Darby begins above the plant as a 10-inch-deep riffly stream, but below the plant's outfalls, the mean channel depth is about 6 feet. The tremendous pulses of floodwater that come off the plant have had precisely the same effect on the streambed as a mechanical dredge, and the Flat Branch has essentially been hydrologically channelized. And this may be only the beginning, for the Honda plant has become the nucleus for a wave of industrial development. A recent expansion of the highway that connects the plant to Columbus not only added 600 acres of impervious surfaces, but also prompted a change in zoning along the highway corridor from "agricultural" to "light industrial." The headwaters of the Darby are on the brink of permanent degradation, for the entire area is headed for 20 percent or higher impervious surfaces—and pavement once laid is almost never removed.

There is a rather straightforward solution to the stormwater problem. If, instead of dumping stormwater directly into the nearest stream, a development were to route its runoff into a series of settling ponds and artificial wetlands that stored the water and released it slowly, channelization and sedimentation would largely no longer be an issue. The Nature Conservancy has been working with the Regional Planning Commission and the City of Columbus to incorporate such requirements into development regulations, and as a consequence, a number of new developments have built settling ponds as a part of their stormwater management plans. The difficult part, though, is generating the political will to make such measures mandatory. Ponds and wetlands take up a lot of space, and in suburbanizing areas where land prices are high, undeveloped land within a subdivision represents a significant loss of income for a developer.

And effective stormwater management is more than just digging a pond at the end of the pipe. If all that is required of a developer is to build the physical structures and walk away, it won't be long before all a settling pond does is delay the impact by the couple of hours that it takes the pond to fill up. Settling ponds and wetlands require careful design and ongoing maintenance if they are to do their job well. The real challenge will not be getting these systems built, but rather creating an effective and enduring social mechanism that keeps them working.


At the far end of a small dirt yard behind the aquatic ecology building at Ohio State University, Tom Watters has set up a series of huge plastic tanks that for all the world look like abandoned hot tubs. Each is full of water and is rigged up with a bubbler, but there are leaves floating on the surface—and 5-gallon buckets full of gravel at the bottom. These tanks are in fact one of several repositories for native unionids that have been pulled out of the Ohio River and other zebra mussel–infested waters in a last-ditch attempt to forestall the imminent extinction of dozens of species of pearly mussels. In sites where native mussel numbers are plummeting, Patricia Morrison and other researchers have for several years been culling out any remaining live individuals and sending them to various holding facilities, including Dr. Watters's tanks here in downtown Columbus. At least in the short run, there is really nothing else they can do.

Their intent at this point is simply to keep some mussels alive, but even this most basic of goals has proven elusive for some species. A group of unionids called the heel-splitters has shown high mortality in captivity, perhaps because they are unable to close their shells completely and consequently desiccate when removed from the water. Other species seem to tolerate life in a tank. But Dr. Watters has not yet attempted captive propagation of even these relatively hardy species—let alone the ones he can't even get to survive out of the river. (Other researchers have had some success at captive propagation, but not at anything like the rates that would be needed to sustain a viable population.) These mussels are in a holding pattern, and no one is sure just how long they will survive.

But even a successful captive propagation program wouldn't really help these native mussels in the long run. The real problem is that the species native to the Ohio and all the other habitats invaded by zebra mussels now have no place to go. "We've got all these species in here," Dr. Watters said. "Now what are we going to do with them?"

De-engineering the Ohio is worth a try. If the Corps of Engineers can modify its headwaters release schedule to simulate natural floods and alter mainstem dam structure to restore some habitat, native mussels might be able to reinhabit portions of the river. But if the habitat these mussels need to survive proves beyond restoration, some of the mussels in these tanks may soon join the passenger pigeon as species whose last members expire long after the species itself has become functionally extinct.

The unionids that are most likely to survive are those that are lucky enough to inhabit a place—such as the Darby—that humans have somehow managed to leave alone. In a world where invasions are becoming ever more frequent, human degradation of habitat invites further degradation by invasive nonnative species. This is why the work of the Darby Partners is so important. Saving high-quality natural areas can't provide immunity to invasion, but it is often the very best kind of protection that people can offer.

CHAPTER 5

# Rolling the Ecological Dice

*Invasiveness, Invasibility, and the*
*Ecological Consequences of Invasion*

## The Damaging Minority

Introducing new species is an ecological version of Russian roulette, but with one hundred chambers in the gun. Of a hundred species introduced, only five to fifteen will establish in the wild; of those that establish, only one or two will cause significant harm. Overall, about one in a hundred turns out to be an ecological disaster—the bullet in the chamber.

These damaging invaders are of great ecological importance. The impacts of invasions can sometimes be mitigated with various control methods, but they can only get worse if ignored. Though they are often described as "biological pollution," the effects of invasive nonnative species are in several respects even worse than the effects of chemical contamination.

First of all, chemical contamination can be stopped, and when it ends, the quantity of pollutants in the environment declines. Organic compounds degrade; even elements like lead and mercury are eventually entombed in sediments. But biological pollutants are self-perpetuating. Once established, nonnative species can sustain themselves and their impacts indefinitely, unless they are suppressed by biological control methods.

Second, chemical pollutants become diluted as they are dispersed by water or wind. Minute traces of DDT might reach even the North Pole, but the quantities will be reduced thousandsfold from the concentrations in the originally polluted areas. Nonnative organisms, however, can multiply as they spread, increasing rather than decreasing in concentration. A thousand miles from the point of introduction, a spreading plant or animal may be even more abundant than at its point of origin if local conditions favor its growth.

Finally, just as some chemicals may synergize each other's effects, so too can some nonnative species enhance each other's impacts. But alien species can even increase the likelihood that other invaders will establish. For example, nonnative birds facilitate the introduction of alien weeds into new habitats, as happens in Hawai'i when introduced mynah birds move seeds of the nonnative shrub *Lantana camara* from pastures into native forests.

These features mean that damage to native species and communities from alien species is both cumulative and potentially self-reinforcing. No other kind of pollution caused by humans has this combination of traits. However, since only a small percentage of nonnative species turn out to be invasive, it is very important to avoid confounding origins—native versus alien—with effects. Impacts are what really matter. Three questions can be used to help predict and define impacts, at least in general terms: What features of nonnative species are likely to promote their establishment? What habitats are most prone to invasion and susceptible to damage from invaders? And what kinds of ecological harm can invaders cause?

## The "Good" Invader

Few people do invasion experiments. The timescale and costs are too large, and the dilemma of not being able to "take back" the invader after the experiment is over makes deliberate introductions simply for the sake of experimentation difficult to justify. Rather, invasion potential is inferred from historical records of unplanned arrivals or past deliberate introductions of species assumed to be beneficial.

The arrival of a new species is not in itself a bad thing. Moreover, the characteristics associated with ability to establish are often different than the attributes that allow some nonnative species, once established, to become invasive and do significant damage to native species or communities. But because establishment creates an irrevocable opportunity for a species to cause harm, the characteristics that favor establishment merit examination.

Unfortunately, almost no one has kept the kind of detailed records of releases and escapes of nonnative species that would be necessary for a rigorous evaluation of the factors that promote establishment. There is no incentive for an importer such as a plant nursery to keep track of the species it introduces—and no one is paying attention when a stowaway species escapes and establishes on its own. The only major exception is biological control, because the release of natural enemies for biological control is by definition a planned invasion and scientists generally keep records of releases (see Chapter 11).

However, even these records don't tell much. A study of biological control releases of insects found that the only factor that consistently predicted establishment was what is known as propagule pressure, or the number of individuals released and the number of times and places they were released. A study of birds introduced to New Zealand was only a little more revealing: It indicated that migratory status was also a factor in predicting establishment, in that nonmigratory species were more successful at establishing than were migratory ones. Overall, determining specific causes of establishment is difficult because analysis of often incomplete historical records is inherently a less effective method of obtaining information than is running experiments.

## Factors That Promote Invasiveness

Once a nonnative species becomes established—that is, once it is self-reproducing in the wild—the particular features that allow it to become invasive depend in part on the nature of the habitat that is being invaded. Nonnative plants that invade agricultural settings are probably the most familiar; indeed, many of the most common agricultural invaders—for example, Queen Anne's lace and dandelion—are probably mistaken for native species by many people. Disturbance is the defining characteristic of agricultural habitats, and adaptation to disturbance is therefore the hallmark of species that are invasive in farmers' fields. A study of invasiveness among agricultural weeds found that the most important plant pests are "primarily herbaceous, rapidly reproducing, abiotically dispersed species." Translated, this means that most weeds in farmers' fields are fast-growing grasses and herbs with windborne seeds.

Disturbance is generally not as severe in natural areas as in farmer's fields that are plowed every year. In such places, damaging weeds generally tend to be species that can invade established plant communities rather than species adapted to colonize bare ground. In this setting, effective invaders must be able to spread rapidly and must have some sort of competitive edge that is specific to the habitats they invade. According to the same study, invaders in natural areas are often "aquatic or semi-aquatic [species], grasses, nitrogen-fixers, climbers and clonal trees." Practice has also shown that some shrubs, such as glossy buckthorn and saltcedar, are likely to become invasive in natural areas as well.

Floating plants that become invasive in aquatic habitats often do so because their position at the surface of the water allows them to monopolize the sunshine. *Salvinia molesta*, a floating fern used in the aquarium trade that was released in Papua New Guinea (and which recently has become established in Texas as well), quickly took over lakes and sluggish rivers by forming a thick layer over the entire water surface, thus cutting off light to all submerged plants. Water hyacinth forms similar floating mats throughout the southeastern coastal areas of the United States and on most major lakes and rivers in Africa. Its effects on native species have been particularly dramatic in Lake Victoria in East Africa, which harbors one of the greatest examples of vertebrate evolution in the world, the cichlid fishes. Hundreds of species are disappearing, in part because their color-based mating system breaks down where visibility is impaired by weed mats.

Many grasses are invasive in part because one of the most common kinds of disturbance in natural areas—fire—tends to promote grasses at the expense of other species. Fire may suppress other plants without affecting the invading grass, giving it the edge it needs to take control of the area. Because the roots of grasses are not affected by fire, these species rebound well after a burn. However, many potentially competing plants such as native woody shrubs and trees suffer dramatic declines as a result of the same fire that promotes the invasive grasses.

Plants that can fix their own nitrogen will quickly dominate habitats

The water hyacinth was imported for its beautiful flowers, but has proven a highly damaging pest. (Provided by USDA-ARS, Aquatic Plant Control Research Unit)

A wetland in Florida when it was free of invasive water weeds. (Provided by USDA-ARS, Aquatic Plant Control Research Unit)

The same wetland after invasion by water hyacinth. (Provided by USDA-ARS, Aquatic Plant Control Research Unit)

Kudzu takes over fields, forests, and sometimes even houses in the southeastern United States. (Provided by USDA-ARS, Aquatic Plant Control Research Unit)

with low soil nitrogen levels. In such habitats, native plants are adapted to a lack of nutrients and will not be able to grow as fast as invaders that can fix as much nitrogen as they need. Some of the most ecologically damaging invaders are nitrogen-fixers. For example, the tree *Myrica faya* is able to invade bare mineral soil in new lava flows in Hawai'i, pushing out the many native species that have evolved the ability to survive in soils with low nitrogen levels.

If a plant is able to grow rapidly and can climb up into native trees, it can position itself between their leaves and the sunshine. Also, vines devote relatively little energy to structural support, which allows them to devote more resources to foliage and fast-growing stems. Oriental bittersweet in the northeastern United States and kudzu in the southeast are both aggressive climbers that smother native vegetation.

A clonal growth form is also an indicator of invasiveness, for it spreads the risk borne by any one stem or portion of the plant among all the stems that share a common root system. Such a mutual aid network can enhance the ability of a patch of this type of species to rebound from attacks and to invade patches of other vegetation. Clonal trees like aspen and many ground covers used in landscaping spread in this fashion and are notoriously difficult to eradicate.

Estimating the potential invasiveness of animals is more complicated than for plants. The success of an animal population depends not only on its ability to survive local competition and predation—limitations it shares with plants—but also on existence in the release area of adequate numbers of vulnerable prey. However, a successful animal invader may have no particular competitive advantage over local similar species and still manage to establish if it is able to feed upon a native host or prey that is unable to mount an effective defense. For example, the hemlock woolly adelgid insect has become a major pest of hemlock trees in the eastern United States not because it is more aggressive or more hardy than native insect pests, but rather because the eastern hemlocks have almost no resistance to the adelgid's attack.

## Statistical Analysis of Invasiveness

A more rigorous approach can be useful for predicting the invasiveness of very specific groups of species. Techniques such as multivariate regression and discriminant analyses are used to correlate the tendency of particular species to invade with the specific features of each species in the study group. For example, many species of pines have been planted in areas of the Southern Hemisphere that have no native softwood lumber species. In New Zealand and South Africa, some pines have established seedlings outside of

Some pines planted for forestry in New Zealand have proven invasive in native grasslands. (Pauline Syrett, Landcare Research, New Zealand)

plantations, and statistical analyses have been used to judge the invasion potential of other species of pines already planted in the country.

Analyzing data for twenty-four species of pines planted in Southern Hemisphere locations, University of California researcher Marcel Rejmánek found that invasive pines tended to be those with small seeds that set large seed crops frequently. These traits meant that the seeds of such trees were produced in large quantities and were more easily carried outside the planted area on the wind. Based on these distinctions, it appears that twelve of the pine species used in forestry in the Southern Hemisphere are somewhat *less* likely to be invasive than are others—and that the others might best be phased out of use.

In Australia, a rating system has been developed to judge the potential for invasiveness of plants proposed for introduction from South Africa, a country whose similar climate makes it a major source of imported plants entering Australia. Plants were found to be likely to become agricultural weeds if they were weeds in South Africa or had ranges that spanned many climatic zones. Researchers also found it useful to ascertain if the plant was closely related to a species already known to be a weed, even if the species in question was not a weed itself. However, while this scheme was useful for predicting a plant's possible status as an agricultural weed, no factors could be identified that would predict if a plant would become a pest in natural areas.

The U.S. National Park Service has developed a method to rank potential harm to natural areas by established nonnative species. Its system is based on answers to a series of twenty-three questions about the plant's biology, local occurrence, abundance, and feasibility of control. Such an approach can be used to identify species that should be given priority for control efforts.

In principle, the same methods of prediction apply to animal and pathogen species as to plants. In practice, we know much less about these groups. Success is often strongly influenced by the context of the invasion. One generality that is easy to make is that nothing succeeds so well as introduction into an ecological vacuum. Mammals of all kinds have been extremely successful on remote islands, which often have no native mammals except bats. European rabbits, for example, have successfully established on many oceanic islands, but despite many releases, they have never established in continental Africa, an area with many competing and predacious mammals.

Given that we have only a very general ability to forecast which species will be the bullet in the ecological pistol, extreme caution seems in order when choosing any nonnative species—plant, animal, or anything else—for introduction. Species with risky attributes clearly should be kept out. Even those with few risky attributes are not necessarily safe. Invasion is a tremendously complex process with far too many variables for consistently accurate modeling, and predictions of invasiveness are about as reliable as predictions of the weather. In either context, it's better to play it safe.

## Characteristics of Invadible Habitats

From the tropics to the tundra, almost all habitats—oceans, rivers, ponds, grasslands, forests, agricultural areas, sand dunes, shrublands—are at risk of invasion. A very few habitats such as cave systems or hot vent communities in the ocean depths are probably in little danger because their highly unusual characteristics require extreme specialization, but these are the exception. In general, nonnative species are more likely to establish if they invade habitats with lowered biotic resistance, encounter prey with poorly developed mechanisms of self-defense, or invade habitats unaccompanied by their specialized natural enemies or when their invasion is facilitated by earlier invaders.

HABITATS WITH LOWERED BIOTIC RESISTANCE. People can reduce the numbers of competitors, predators, or pathogens an invader will encounter by changing the invaded habitat. For example, disturbed soil often favors the establishment of invading plants. Tilled agricultural soils or overgrazed grassland soils are common points of invasion for many kinds of weeds, for the absence (or degraded condition) of native species on such abused lands reduces competition that might otherwise prevent new invaders from getting a foot in the door.

In the western United States, nonnative annual grasses have replaced native perennial bunchgrasses over large areas because grazing patterns established by ranchers lowered the competitiveness of native grasses. Most native grasses are adapted to withstand light to moderate grazing only in winter. After humans replaced native grazers with cattle, grazing became not only more intense overall, but also shifted from winter—when perennials are least affected—to spring and summer, when grazing does great damage to perennials but is well tolerated by fall-germinating annuals. Under these circumstances, alien grasses better adapted to the new grazing environment gain an advantage and replace the native grasses.

COMMUNITIES OF UNDEFENDED PREY. In isolated communities, native species lose defenses against threats that are not present. If there are few or no large herbivores, plants tend to lose toxins and thorns; if there are no predators, birds will sometimes lose the ability to fly. This is why almost any vertebrate invading an oceanic island has flourished. For example, rabbits introduced to Laysan (one of the low Hawaiian Islands) exterminated a large portion of the island's mostly endemic flora in only twenty years. The native plants had few defenses against mammalian grazers, for they had evolved in the absence of any grazing pressure. Similarly, the brown tree snake easily decimated the forest birds of Guam when it invaded in the 1950s because these birds had never before been confronted with any predacious snakes. Isolated species lose defenses against nonexistent threats, leaving the ecosystems they inhabit open to invasion by nonnative species that introduce such threats.

HABITATS WITHOUT AN INVADER'S SPECIALIZED NATURAL ENEMIES. When a species is moved from its native habitat to a new ecosystem, it can sometimes escape the specialized herbivores, predators, or pathogens that controlled its numbers in its native habitat. For example, purple loosestrife is a Eurasian plant that has become highly invasive in North American wetlands. Its foliage- and root-feeding insects were left behind in Europe when its seed came over in ballast soil of sailing ships at the beginning of the nineteenth century. Although it displaces cattail and other native plants in the wetlands it invades, loosestrife does not appear to have any intrinsic features of growth form or position in the habitat that give it a definite advantage over cattails. Rather, it seems that escape from its attacking insects and diseases has allowed it to redirect energy into growth that would otherwise have to be used for defense or regrowth after attack (see Chapter 11).

HABITATS WITH ESTABLISHED INVADERS THAT FACILITATE NEW INVASIONS. Previously introduced nonnative species may themselves cause habitat disturbances or lower the competitiveness of native species in ways that favor more invasions. As was discussed in Chapter 1, pigs and strawberry guava—both invasive nonnative species in Hawai'i—promote each other and synergize the damage they do to Hawaii's native forests. Pigs make seedbeds in the forest by rooting the soil and then deposit feces containing guava seeds; the guava trees that result provide food for yet more pigs. As stands of strawberry guava mature, they dominate the former disturbed areas, inhibiting native forest regeneration. With the help of pigs, strawberry guava spreads deeper into native forests and does more damage than it would on its own.

## Ecological Consequences of Invasions

Established nonnative species affect native species and ecosystems by changing the rules of the food web in the invaded community. Some changes affect only one layer in the food web by changing how organisms compete for resources. For instance, invasive plants sometimes cause *generalized super-competition* in native ecosystems, dispossessing entire native plant communities of the space, light, or other resources they need to survive. Invasive animals that exert a competitive effect are usually (though not always) more specific in their impacts, replacing a single similar native species or group of species in what could be called *niche usurpation*.

However, many animal invaders make themselves felt not so much through competition as through exploitation of organisms at a lower level of the food web. Feral sheep and other vertebrate herbivores devastate islands through a process of *generalized superexploitation*, eating a wide variety of native plants to near extinction; predators such as the brown tree snake have similar effects on some groups of island birds. Other invasive animals (generally insects) and most nonnative pathogens are more specific in their depredations, exploiting only one or a few target native species in a kind of *specialized superexploitation*.

Finally, some invasive species can affect the very basis for the entire food web, shifting the dynamics of an ecosystem to a fundamentally new pattern. Invasions characterized by this kind of *habitat modification* can be among the most catastrophic and intractable of nonnative species problems, for major alterations of the soil chemistry, hydrology, or disturbance regime of an ecosystem often cause such massive damage to native communities that recovery can be difficult or impossible even if the invader is controlled.

## Generalized Supercompetition

When a species is moved to a new habitat far from its native home, the biotic forces that played a role in slowing the rate of growth of the invading species' population are often left behind. This does not necessarily mean that the species will become a damaging invader in its new habitat, or even that it will manage to establish itself there; but replacement of such strong, specialized natural counterbalances with less specialized ones enormously increases the chances of a biotic explosion in the invaded area.

Both islands and continents are vulnerable to damage from invasions of such supercompetitors. Garlic mustard is a European woodland plant that forms dense stands beneath deciduous forests in the U.S. Midwest, Northeast, and Southeast, virtually eliminating native herbaceous plants from invaded sites. Like many other invasive nonnative plants—among them Japanese knotweed, buckthorn, and purple loosestrife—garlic mustard uses a "blanket coverage" strategy to take control of invaded areas.

For most native plants, dispossession of their habitat is the end. In the fynbos of the Cape Town region of South Africa, native species are being overrun by nonnative woody shrubs, especially species of *Hakea* and *Acacia*. Many of the native plants cannot grow in areas dominated by these invaders. An ambitious program of cutting and herbicide control of these supercompetitors has made significant inroads into infested areas, but such solutions deprive invaders of their competitive advantage only temporarily. In the long run, the complex ecological surgery of biological control—an uncertain but tremendously powerful tool for ecological restoration—may be the only method that has any real hope to check such losses (see Chapter 11).

## Niche Usurpation

Another outcome of species mixing is simple competition between functionally similar species for possession of a single niche. This is a specialized form of supercompetition that can decimate or exterminate specific species. For example, zebra mussels displace native mussels by appropriating the native mussels' food resources and by blanketing the substrate they need to live. Rainbow trout stocked in streams outside their range can displace local endemic trout. In such cases, the competing species are similar in type and differ mainly in the details of their biology. Impacts of such introductions are relatively specific in that the nonnative species affects a single species or a group of closely related species.

One mode of displacement involves the takeover of breeding sites of native species. Introduced starlings severely reduced bluebird populations in North America by occupying the cavities in dead trees that bluebirds need as nesting sites. Bluebird numbers rebounded following a nationwide program to encourage the placement of artificial nesting boxes with holes just big enough for bluebirds to enter—but too small for starlings.

Invaders can also take over the niches of native species by preempting the food resources that the native species need to survive. A few highly competitive "tramp" ant species such as the Argentine ant, big-headed ant, and crazy ant that move easily with human cargo are displacing native ants around the world. Crazy ants, brought to Colombia by foresters to "get rid of snakes," have instead eliminated thirty-six species of ants—95 *percent* of the local species—through aggressive competition for resources.

In the aquatic realm, the American bullfrog—an eastern species—has caused a 90 percent reduction in populations of the local native yellow-legged frog in California through competition between the tadpoles of the two species. Political boundaries sometimes obscure the real geography of ecology, for although bullfrogs are native to the United States as a political jurisdiction, they are not native to California as an ecological region. Similarly, the rusty crawdad is another native U.S. species that is outcompeting similar species in habitats where it is introduced. Its sale as fishing bait has led to releases in many ponds and rivers where it has displaced local native species of crawdads, many of which are limited-range species in danger of extinction.

When native animals are displaced by invaders, biologists and amateur naturalists may be the only ones who notice. Unlike supercompetitive plant invaders whose thickets and mats are impossible to ignore, the replacement of one kind of frog with another is subtle and easily overlooked. But the net result is a loss of species diversity as many local species are replaced by one or a few highly competitive types across a large region.

## Generalized Superexploitation

Biological invasions can also cause dramatic changes in native communities when generalist superexploiters decimate whole groups of native species at the next level down in the food web. Among species that have a long-standing association, relationships between the various parts of the food web usually maintain a rough equilibrium over time. However, if a new herbivore or predator joins the system, it may become a superexploiter in the invaded community.

Island species are more likely to be particularly vulnerable to this kind of damage because they are relatively more isolated—and thus unprepared for a wider variety of exploiters—than are continental species. However, even continents are isolated at the macrolevel of biogeographical provinces, and continental species are vulnerable to invaders from different ecological regions. The key features of these and most other instances of superexploitation by invasive species are that the native species in question were evolutionarily unprepared to tolerate the exploiter's attacks, and the community lacked other species able to suppress the exploiter. Unlike cases of supercompetition, in which invasion generally causes ecological degradation and population losses but not complete elimination of native species, superexploitation can lead to extinction of the species under attack.

When the Nile perch was moved from the Nile River drainage into Lake Victoria in East Africa, this large voracious predator ate its way through one of the most complex and interesting fish communities in the world. Faced with an attack against which they had virtually no defense, most of the lakes's native cichlid fishes simply disappeared, and some two hundred species found nowhere else in the world may now be extinct. The food chain collapsed and underwent massive reorganization, and a biting midge that formerly was rare became a severe problem. By introducing a highly efficient nonnative generalist exploiter to the lake, a government program to "enhance" a fishery has caused a greater loss to the vertebrate diversity of the continent than have all the poachers in Africa.

## Specialized Superexploitation

If a specialized invader encounters prey that have little or no evolved resistance to its particular mode of attack, or if the invading species leaves its controlling natural enemies behind, it can sometimes exploit its target species with extreme intensity. The forests of the eastern United States have suffered many losses to pests of this kind, mostly pathogens and specialized invasive insects. Chestnut was decimated in the first half of the twentieth century by chestnut blight. Dutch elm disease has killed many of the elms that once lined city streets. Butternut is now exceedingly rare because of an introduced canker. The American beech is dying over most of eastern North America from the combined impact of a nonnative scale insect and a complex of virulent pathogens. These and other pathogens and pests move nearly invisibly on nursery stock, plant specimens sent to botanical gardens, or raw wood imported from overseas. Sometimes the pathogens don't even cause any obvious damage to the host plant species on which they arrive, but turn out to be lethal to a native species in the invaded region. Few alien species have greater impact than these pests and diseases—and few are harder to stop (see Chapter 6).

## Habitat Modification

Most alien species don't modify the basic ecological dynamics of the places they invade. The few that do can alter native communities profoundly. Invasive species of this kind can change the frequency or duration of natural fire, raise or lower the local water table, and alter the soil's fertility or chemistry. Such habitat changes create shifts in the fundamental conditions within which native species must live.

Fire frequency and intensity vary widely across different kinds of natural communities. Some almost never burn; others (such as the California chaparral) require frequent fire to retain their characteristic composition. Invading plants can either increase or decrease the frequency or intensity of fires. For example, when nonnative grasses invade natural habitats, fire frequency may increase, reducing the diversity of native plants not adapted to frequent fire. In Hawai'i Volcanoes National Park, invasive grasses have caused increased fire frequency in woodland communities, leading to a reduction in numbers of native plant species. In areas burned twice within twenty years, native plant diversity declined 84 percent compared to unburned sites. In another case, removal of feral goats from a seasonally dry area in Hawai'i led to dramatic increases in the densities of two invasive, flammable alien plants, broomsedge and bush beardgrass. As fire increased in the goat-free habitat, the native shrub *Styphelia tameiameiae*, which had withstood centuries of overgrazing by the goats, disappeared because it could not support the increased exposure to fire caused by the nonnative plants.

Native plants typically are closely tuned to local water conditions as well, and significant changes—up or down—in the amount of available soil moisture will often render sites unsuitable for their growth. In the southwestern United States, for example, watercourses are centers of biodiversity, and conservation of such habitats is a regional priority. Many such riparian areas are now dominated by saltcedar, a group of related invasive plants imported from Asia in the nineteenth century to stabilize sand dunes that has since spread along creeks and rivers throughout the West. While the dense stands of saltcedar along many southwestern waterways suggest that it is simply a supercompetitor, its full impact on the ecosystem is much more profound than simple crowding. The native plant communities along these creeks are not so much crowded out as dried out. Saltcedar's roots penetrate up to 20 feet into the soil—far deeper than almost any native plant. Since it has constant access to the water table, saltcedar does not have to limit

Saltcedar invades riparian areas in the desert Southwest, where its deep roots can lower the water table beyond the reach of most native plants. (Jason Van Driesche)

the amount of water it uses to survive. As a consequence, thick stands of saltcedar can cause water levels in the soil to drop. When this happens, the characteristic habitats of riparian areas—the moist soils, the wet meadows and seeps—disappear rapidly. With them go the native plants that live only in these habitats.

Even invasive plants that don't burn out or dry out the native vegetation can change the soil in ways that make it impossible for the native plants to compete. The issue is not the quality or chemistry of the native soil per se, for native plants will be adapted to conditions as they are. What cause problems are changes of any sort in basic soil conditions, because any significant alteration will likely put native species that are adapted to historical conditions at a competitive disadvantage. Some species change soil chemistry by synthesizing or concentrating toxins in their leaves and then depositing these chemicals on the soil when the plant sheds its leaves. For example, ice plant causes the soil to become more saline, and many eucalyptus species produce phenolic compounds (the "eucalyptus smell") that are toxic to most other plants. Still other plants change not the quality of the soil, but the quantity, by accelerating or retarding erosion rates. For example, Australian pine increases coastal shoreline erosion in Florida. These changes in soil characteristics alter the fundamental ecological rules of the ecosystem. When the rules change, many native species are unable to compete.


The ecological roles of nonnative species are as varied as the species themselves. This means that documenting the presence or absence of a species whose origins are elsewhere by itself tells very little about the relative importance of such new arrivals or the degree to which they might alter the native ecosystem. Addressing the impacts of nonnative species in a meaningful fashion requires a measure of discrimination, for portraying all alien species as damaging is counterproductive. The most useful kind of defense for ecosystems threatened by invasive species is careful analysis of the very real threats posed by the damaging minority.

# CHAPTER 6

# Fading Forests

## *Invasive Pests and Forest Destruction in Eastern North America*

A species doesn't have to go extinct to become an ecological ghost. The first European colonists encountered hundreds of kinds of trees in the temperate forests of eastern North America, and every one of these species still exists. But a number of trees once abundant in these forests have been reduced to shadows—nominally present, but functionally gone.

The American chestnut was once the dominant tree across tens of millions of acres. Now it is nothing but an understory shrub sprouting and dying from the roots of toppled giants. The American elm, once one of the best-loved shade trees in the nation, is now quite rare. Flowering dogwoods that once lit up the forest understory throughout the mid-Atlantic region are now mostly blackened and dying. Butternut, once a widespread hardwood, is now a candidate for listing as a federally endangered species. The once-smooth bark of the American beech is now cracked and gnarled across much of the species' range. In these and many other cases, the principal culprit is an invasive nonnative pest or disease that has proven more devastating to its adopted North American hosts than to the related tree species that the pest attacked in its native range.

Where did all these invaders come from? How were they introduced to North America? Why are they so devastating? Though the answers to these questions are different for every species, an in-depth examination of the origins and spread of a few major invaders provides a degree of insight into the problem as a whole. Ultimately, it is the cumulative impacts of all these invaders that is most worrisome, for while the destruction of any particular

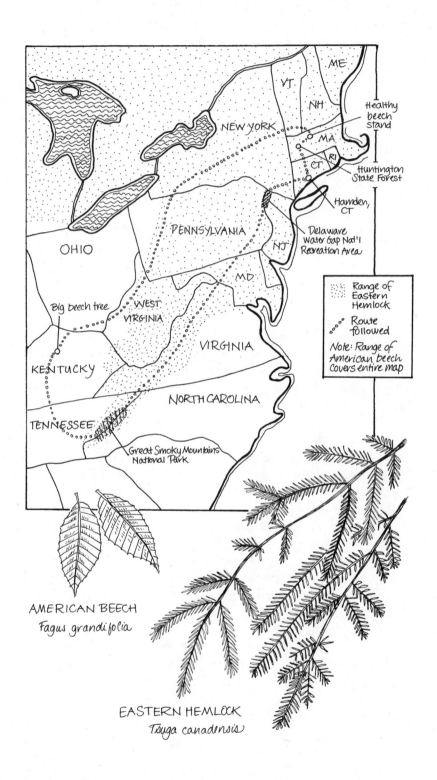

ME

VT.

NH

NEW YORK

Healthy
beech
stand

MA

RI

CT

Huntington
State Forest

Hamden,
CT

PENNSYLVANIA

NJ

Delaware
Water Gap Nat'l
Recreation Area

OHIO

MD.

Big beech tree

WEST
VIRGINIA

VIRGINIA

KENTUCKY

NORTH CAROLINA

TENNESSEE

Great Smoky Mountains
National Park

Range of
Eastern
Hemlock

Route
followed

Note: Range of
American beech
covers entire map

AMERICAN BEECH
Fagus grandifolia

EASTERN HEMLOCK
Tsuga canadensis

tree species is a genuine loss, the fading of the forest as a whole may well lead to catastrophe.

## A Complex and Maddening Disease

The forest in which we walked was a natural community common to southern New England, dominated by American beech and sugar maple with a variety of other hardwoods and hemlocks scattered throughout. But Dr. Dave Houston, the U.S. Forest Service plant pathologist who met me this morning at this state forest in western Massachusetts, was singularly focused on the beeches. They are disfigured and dying, victims of an insect-fungus complex known as beech bark disease. In studies across eastern Canada and the northeastern United States, Dr. Houston has elucidated much of what we know about this disease. But even for him, it seems there are more questions than answers.

As soon as we entered the forest, Dr. Houston pulled out his magnifying glass. The beeches were covered with tiny white specks. "Beech scale," Dr. Houston said as he moved in for a closer look. These flecks of white are the nonnative insect component of the complex of species that collectively causes beech bark disease. The scales look more like bits of lint than insects because what we see is the wax that covers their bodies in protective woolly strands. If conditions are favorable, Dr. Houston explained, the scale can become so numerous that beech trunks turn from gray to white. That happened around here in 1984, when a dry fall and a mild winter the year before increased scale survival and gave the population a head start in the spring. "Nobody remembers seeing it like that before," he said. A lot of beeches died in the couple of years that followed.

These trees died not from the presence of the scale alone, but from one or both of two fungi (one introduced and one native, but both in the same genus) that take advantage of changes in the bark created by the scale. These changes allow the fungus to get past the tree's defenses and infect and kill the cambium layer. Dr. Houston took out his penknife as we walked and carved away the bark on another diseased beech to reveal the damage beneath. "Look over here," he exclaimed. "You can really see the damage." The inner bark was shot through with red. The tree was heavily infected with at least one of the *Nectria* fungi and probably didn't have long to live.

But many other trees in this same stand have had the disease for years and manage to hang on, scarred and stunted but alive. Still others show complete immunity to the disease and stand tall and smooth-barked among their infected neighbors. This spectrum of response is part of what makes the impact of beech bark disease so complicated, but it also gives hope for the future of the species. Foresters and ecologists from Nova Scotia to West Virginia are faced with forests filled with beeches that refuse to die, but that are too far from health to do much more than take up space. But there is hope to be found in the fact that a few trees seem to be unaffected by the disease—a fact that suggests that given enough time to understand the dynamics of a diseased stand, Dr. Houston and his colleagues might one day learn how to steer its makeup toward immunity.

That day is still distant, though, and in the meantime the disease continues its southward spread. The scale was accidentally introduced to the Halifax Public Gardens in Nova Scotia near the end of the nineteenth century. Like many of the early forest pests, beech scale came in as a hitchhiker on ornamental nursery stock. It wasn't noticed until 1911, by which point it had already spread to native beech in nearby Bedford, Nova Scotia. Presumably, the nonnative species of *Nectria* fungus soon followed the scale, and the two moved across New Brunswick and Maine more or less in tandem over the course of the first half of the twentieth century. By 1975, the disease had spread across New York and into northern Pennsylvania, causing severe damage to beech stands across the northeastern United States. By the early 1990s, infestations were well established in West Virginia, northeastern Ohio, and even in the Smoky Mountains along the North Carolina–Tennessee border.

What Dr. Houston and I were examining is an "aftermath" forest. The scale and the introduced fungus invaded this forest in Massachusetts sometime in the 1950s. The most susceptible trees are long gone, killed probably within five or ten years of first infestation, and most of the remaining trees are battling the disease with varying degrees of success. Anyone with even a passing familiarity with beech, though, would see how sick these trees are. Their trunks, normally smooth even on the largest specimens, look more like the furrowed boles of old sugar maples. Many show thinning crowns, dead branches, and yellowish leaves—all sure signs of a tree in severe distress. Beech trees continue to die in this forest at a pace that far exceeds their mortality rate before the killing front of the disease arrived.

Farther down the trail we came across a tree that seems happily out of place among its scarred and twisted neighbors. It appeared to be among the fewer than one in a hundred beeches that are immune to the disease. What distinguishes these trees from those that sicken and die? Dr. Houston believes it may have something to do with the level of nitrogen present in the bark. The bark of immune trees has a lower nitrogen content than does the uninfested bark of susceptible trees. It seems that the scale is unable to

Beech bark disease turns the American beech's smooth gray bark into a rough mosaic of cracks and pockmarks and often kills the tree. (Jason Van Driesche)

establish on the nutritionally deficient low-nitrogen bark of resistant trees. And without an insect infestation, the fungi simply never have an opportunity to enter and kill the inner bark.

The obvious question, then, is why not cut down the diseased and dying trees, and let the immune individuals repopulate the forest? Dr. Houston smiled and shook his head. "That's if we knew how to do it," he said. Unfortunately, it's not that simple. Beeches—even diseased ones—can send up vigorous shoots wherever their root systems are injured. Cut down a dozen infected beeches, and a year later you may have a couple hundred susceptible sprouts. And treating the cut stumps with herbicides to keep them from sprouting can backfire. Beech roots graft to one another readily, and anything you use to treat one tree is likely to find its way into the root systems of its neighbors. "So we have to be aware of all these interactions," Dr. Houston explained. Very little about this disease, it seems, is straightforward or easy.

By the early 1990s, it had become clear that very careful cutting and herbicide treatment of only those diseased beeches that were far away from patches of resistant trees were really the only avenues for control that held any promise at all. Chemicals do not work for large-scale control of forest diseases, for it is impossible (and highly undesirable) to spray an entire forest with fungicides. And unfortunately, the biological control candidates found thus far have turned out to be dead ends. There are a few insect predators (such as the colorfully named twice-stabbed ladybird beetle) that feed on the scale, but they do not reduce its abundance enough to have any real effect on the degree to which the fungi are able to enter the bark. The *Nectria* fungi themselves are attacked by another fungus, but the attacker and the disease seem to live quite happily together. While complexity creates possibilities, it provides no guarantees, and saving the beech from this disease is as formidable a challenge as understanding the disease itself.

## The Unanticipated Catastrophe

There isn't supposed to be a view from the road that winds up East Rock. This ridge in southern Connecticut used to be blanketed in eastern hemlocks, and all that was visible from the road was layer upon layer of soft green branches. Now most of the hemlocks are dead. A thicket of black birch has sprung up in their place, and it has yet to grow high enough to cut off the view of the valley below. There isn't much up here anymore that could be called forest.

Dr. Mike Montgomery, an entomologist with the U.S. Forest Service, has been watching the hemlocks die since the late 1980s. They are under attack by the hemlock woolly adelgid, a tiny white insect native to Asia that for many years was thought to be of no real consequence as a forest pest. It had been in North America for more than sixty years and on the East Coast since the early 1950s, and no one had ever seen it attacking anything but cultivated hemlocks. But by the late 1980s, reports began surfacing from across Virginia—the area to which it was first introduced on the East Coast—of tremendous damage and even mortality in forest stands of hemlock. By the mid-1990s, with infestations from northern North Carolina to eastern Massachusetts, the adelgid had become a major problem.

It seems odd that until just a few years ago nobody paid any attention to the adelgid. If it took thirty years to reach pest status on eastern hemlocks, I asked Dr. Montgomery, what's happening to the hemlock species out West? After all, the adelgid has been present there for over twice that long. Shouldn't we have realized a lot earlier that we had a potential problem on our hands?

No one had any reason to believe it would become a pest in the East, he explained, because the hemlock species of western North America are little affected by the adelgid. Everyone simply assumed that the eastern hemlocks would fare the same. But ecological divisions often do not follow political lines, and the hemlock species found in Connecticut are not as closely related to those in Oregon as one might think. Until the Bering land bridge was flooded at the end of the last ice age, the forests of western North America and eastern Siberia were relatively contiguous. The great expanse of prairie and desert that separates the western North American forests from those in the East, however, dates back millions of years to the rise of the Rocky Mountains. The eastern hemlocks have evolved largely in isolation from both the western species and their Asian counterparts and thus have little resistance to the adelgid. This means that even the most catastrophic invasions are not necessarily predictable, for a pest problem is defined as much by the characteristics of the invaded ecosystem as by the attributes of the invader itself.

The greater susceptibility of eastern hemlocks appears to be the key factor in their high level of mortality when attacked by this adelgid. Not only does the insect weaken the tree by depleting it of nutrients, but there is also some speculation that the insect's spittle may be toxic to both eastern hemlock and its other host species, the Carolina hemlock. Their susceptibility is compounded by the fact that the adelgid left all its natural enemies behind when it was introduced to North America. Consequently, the vast majority of infested trees die within five years. And even if the adelgid doesn't kill all the trees in infested stands, it almost invariably eliminates hemlock as a unique community. Surviving hemlocks have a much thinner canopy than healthy trees and do not provide the deep shade and cool temperatures that are unique to a healthy stand of hemlocks.

On the way back to Dr. Montgomery's office, he and I stopped to inspect a stand of heavily infested hemlocks along a reservoir below East Rock. Each twig was traced in white, with an adelgid or two at the base of nearly every needle. The trees' foliage was yellowing and their crowns thin. They looked like discarded Christmas trees at the end of the first week of January, and they didn't have long to live. These trees will not resprout from the roots, for unlike hardwoods, hemlocks keep most of their energy reserves in their needles. Once a hemlock has lost a substantial portion of its foliage, its long-term prognosis is very poor.

Right next to the dying trees, though, was a row of healthy green hemlocks. Unfortunately, they were not immune to the adelgid. These trees were healthy only because they had been sprayed with insecticide, a treatment

The hemlock woolly adelgid insect sucks the sap of eastern and Carolina hemlocks, killing nearly nine out of ten infested trees within five years. (M. Montgomery, U.S. Forest Service)

that is practical and cost-effective only for individual ornamental trees or small stands with particular value. The trees have to be completely soaked at least once a year, and the ecological and economic costs (not to mention the logistical difficulties) of using chemicals on entire stands would be too great. In the long run, the only real hope for saving hemlock as a forest species lies in developing an effective program of biological control (see Chapter 11).

Dr. Montgomery admits this won't be easy, particularly for an insect like the hemlock woolly adelgid. The difficulty stems in part from the fact that there are no known parasitoids of adelgids. (Parasitoids are insects that lay their eggs in or on another species of insect and then kill their hosts when the eggs hatch and the larvae eat their host.) Parasitoids are generally the safest and most effective biological control agents, for they often have high reproductive rates and are able to multiply rapidly in response to sudden outbreaks in their host population. In addition, most attack only a few closely related host species, which keeps their impact focused on the non-native pest they are introduced to control. Adelgid biological control programs must rely on predators, which generally are not as effective at keeping pace with rapid increases in prey populations as are parasitoids. It is also somewhat more difficult to find a predator that is specific enough in its feed-ing preferences that it is both effective and safe enough to use as a biologi-cal control agent.

But there are possibilities, and Dr. Montgomery and several other researchers are pursuing them vigorously. As of 1999, their work had identi-fied several predatory ladybeetles that feed specifically on adelgids, and they had done controlled field tests on one species called *Scymnus sinuanodulus*. What they found was only partly encouraging. While the beetle reduced adelgid numbers by two-thirds in caged trials, it reproduces so slowly in the wild that it will be effective only for keeping low populations of adelgids sup-pressed—not for controlling full-scale outbreaks. The key to control is estab-lishing a group of natural enemies that feed on the adelgid in complemen-tary ways. "I've seen multiple species of beetles, flies, and lacewings feeding on the hemlock woolly adelgid in China," Dr. Montgomery explained, "so we shouldn't expect just one species to control it here."

This makes biological control a difficult game to win. Only those natural enemies that are likely to have no significant impacts on native species should be introduced, a hurdle that can eliminate otherwise effective natu-ral enemies from consideration because they are too general in their feeding habits. Even among those that pass tests for safety, many introduced natural enemies fail to establish because of differences in environmental conditions or lack of long-term support for the project. But when enough carefully eval-uated biological control agents establish and spread, they can be a powerful and precise tool for controlling what are otherwise unmanageable nonnative pest problems—and hemlock woolly adelgid appears to be this kind of pest.

Knowledge of the origins and biology of beech bark disease and hemlock woolly adelgid is only the first step toward understanding the ecological implications of their presence on this continent. Equally important is a broad sense of their impacts across the forested landscape of eastern North America. The consequences of their spread are clearest along the slow-burning lines of each invader's advancing front, in places where healthy trees are abundant and time is about to run out.

## Exploring the Killing Front

Rich Evans lights up when he walks under hemlocks. "Even before I knew anything about forest ecology, when I came into a hemlock ravine, I knew it was different," he said. "You can just sense that it's ecologically different." We were hiking along Adams Creek, a heavily wooded hemlock ravine in the Delaware Water Gap National Recreation Area (a unit of the National

Park system that follows the Delaware River along the border between New Jersey and Pennsylvania). On a muggy day in late June, the air along the creek was soft and cool. The canopy was so thick that the sky was visible only in slivers. Mosses and ferns grew scattered in the dim light of the forest floor. My senses were telling me the same thing Rich saw when he first came hiking here: There is nothing in this part of the world quite like a hemlock stand.

But I was here for the specifics. What species are closely associated with hemlock stands? What specific environmental conditions do hemlocks create or perpetuate? How do these conditions affect the composition and function of the community? And in what ways would the loss of hemlock reverberate throughout the community as a whole? Rich Evans is an ecologist at the Delaware Water Gap and has spent the last six years putting together a coherent picture of how the hemlock ecosystem works. While he stressed that the picture changes somewhat from place to place, it quickly became clear that there are a few basic characteristics of hemlock stands that strongly shape their character and composition regardless of context.

First, a mature hemlock stand creates an extremely dense canopy of dark, evergreen foliage. This has two major effects: It provides an environment that a number of native bird species strongly prefer as nesting cover, and it casts the deepest shade of any forest type in the Northeast. Research at the Gap has demonstrated that at least five species of songbirds are strongly associated with hemlock, and many other birds use hemlock to a lesser degree. And at the level of the forest floor, the deep shade and the cool, moist soil conditions the shade creates promote the growth of a tremendous variety of mosses and liverworts—a total of 123 species in a 1995 survey of two hemlock creeks along the Gap.

Second, streams lined with hemlocks tend to have a more even flow of water than do those in hardwood stands. This combined with the cooling effects of deep shade (the temperature of streams flowing through hemlock stands drops by five to ten degrees Fahrenheit) makes hemlock streams ideal habitat for native brook trout, a species that is often at a competitive disadvantage with introduced German brown trout in larger, warmer streams. A study conducted for the Delaware Water Gap by the USGS Biological Resources Division found that "brook trout were twice as abundant in hemlock streams as in hardwood streams." These same conditions allow a wide variety of aquatic macroinvertebrates such as crayfish and creek-dwelling insects to flourish in hemlock streams as well. There are three species of aquatic macroinvertebrates in the Delaware Water Gap that are unique to hemlock streams and fifteen that are strongly associated with hemlocks; in contrast, no species are found only in hardwood streams.

One of the most effective ways to gauge the magnitude of the changes that would follow the loss of hemlocks is to compare the composition of hemlock stands with that of topographically similar hardwood stands. If a given hemlock-dominated site has the same soil type, slope, aspect, and the like as some particular site that is dominated by hardwoods, then the difference between the two is at least a rough indicator of the kinds of changes and losses a hemlock-dependent community would undergo as hemlocks died off. Using a computerized geographic information system to select pairs of sites, researchers at the Delaware Water Gap did exactly such a study in the mid-1990s. Their findings confirmed that significant differences in flora and fauna existed between the hemlock sites and their hardwood counterparts—and suggested that some of what would be lost with the hemlock stands was unique and irreplaceable. In a report for the 1996 North American Forest Insect Work Conference, Rich Evans stated the likely consequences in more concrete terms: population declines among birds that depend on hemlock, loss of native brook trout, loss of some species of mosses

Hemlocks killed by the hemlock woolly adelgid leave gaps in the forest canopy that cause a major shift in forest community type. (Courtesy of Richard A. Evans, National Park Service)

and liverworts, and increased occurrence of invasive nonnative plants. In short, Adams Creek would lose many of the things that make it, as Rich put it, "different."

This means that the impacts of the adelgid go well beyond the loss of hemlock itself. At the southern end of the Delaware Water Gap, where established adelgid infestations were already killing hemlock stands, Rich has seen aggressive nonnative weeds and trees taking over the openings around dead and fallen hemlock trees. Hemlock holds this ecosystem together, and when it is lost, the ecosystem falls apart.

There are a few large stands of old growth hemlock forest left far to the south of the Gap in the southern Appalachian Mountains of North Carolina and Tennessee. In these old forests, the particular character and the tremendous vulnerability of hemlocks are clearer than anywhere else. Most of the old growth is in the Great Smoky Mountains National Park, at the southern terminus of the Blue Ridge Parkway. In the valley of Cattaloochee Creek, a small drainage at the eastern end of the park, there are groves of hemlock and yellow poplar with trunks four and five feet thick. They possess the clear soaring quality and the air of competence particular to very old trees. These trees are not immortal. But from the ground, it is easy to imagine they are.

But no amount of time, not even five hundred years, can prepare these

In the old growth forests of the Great Smoky Mountains, hemlocks trunks can be 5 feet thick. (Great Smoky Mountains NP)

trees for the adelgid's arrival. Not too many years from now, it will blow in on a storm, or a bird will carry the insect in on its feet or its plumage. Even a tourist driving down the Blue Ridge with a bit of infested branch stuck in a car door or wheelwell could bring it in. There are no apparent impediments other than time and chance to the adelgid's southward spread. It is almost inevitable that the adelgid will eventually make its way to this forest. I stood among those giants and tried not to imagine the future.

Chances are good no one will notice the adelgid for a year or two after it arrives. It will feed and multiply and spread quietly through the canopy. But before long it will become so numerous as to attract attention to itself. Subtly, the light at the forest floor will change. The canopy will begin to whiten and thin. The understory hemlocks will drop their needles and die. Within a decade of infestation, the soft light of this ancient forest will become hard-edged and bright, and all that will be left of the hemlocks will be enormous blighted trunks and a scattering of sick and dying trees.

## Searching for Healthy Beeches

The eastern hemlock is not the only tree that grows to magnificent proportions in the Smokies. There are healthy American beeches with trunks nearly 3 feet thick along the Cattaloochee trail as well. Their smooth, bright bark shines through the softer greens and browns of oaks and hemlocks. Though they have never been particularly common here, beeches once were found from the hollows to the peaks. But by the time I visited the Smokies, the low-lying hollows and valleys were the only place where healthy beeches could still be found.

Until a few years before, American beeches also grew by the thousands high on the shoulders of the tallest peaks of the range in sharply delineated stands known as "beech gaps," so called because they formed islands or bands within a sea of spruce and fir. There were no species unique to the beech gaps, and at something less than 20,000 acres, their aggregate acreage was quite small. But even so, these beech stands did not go unnoticed. If you had asked a hiker along the Appalachian Trail about the beeches in the Smokies, you would likely have been told of translucent green shade on a hot summer day's hike along the ridge. If you were to ask a bear, she would no doubt have praised the crop of nuts that add to her winter fat stores. (Turkeys, foxes, grouse, blue jays, and many others who frequent higher elevations would have joined in.) If you had asked an ecologist, you would have learned of the value of beech in stabilizing high-elevation soils. These beech gaps, small and unassuming though they may have been, were an important part of the local ecosystem.

Beech bark disease is killing beeches on the peaks of the Smokies, eliminating the unique habitat type known as beech gaps. (Jason Van Driesche)

Glenn Taylor is a forestry technician with the park, and his job brings him often to these peaks and gaps. On a hot day late in summer, he and I followed a trail from a parking lot at the crest of the cross-park road over a knob to join another trail on the far side of the ridge that led to several of the beech gaps. As we came down off the ridge, the canopy opened, and we left the trail once again to clamber upward. We were chest-deep in ferns and wildflowers and August sunshine. I stopped to look back and realized that this clearing in the forest was a new one. All around us were dead and dying beeches. What had once been a beech-filled "gap" in the sweep of evergreens that cover these peaks was now a true gap in the forest canopy. Beech bark disease had arrived.

North and west of the park, beech becomes healthy and abundant again in the forests of the southern Cumberland Plateau. But in northern Kentucky and into Ohio—well to the west of the killing front of the disease—they are quite scarce once more. This region was once the heart of the range of the American beech, and on its fertile soils they grew to enormous size.

A healthy beech stand like this one in central Massachusetts is a rare sight. (Jason Van Driesche)

But in the Bluegrass region of northern Kentucky and the farmland of Ohio, almost all the land was long ago converted to pasture and crops. There is very little forest, period.

In the Bluegrass, the only forest stands that persist are found along rivers and streams, and it was along the Kentucky River that I came across the biggest beech I have ever seen. Its girth was greater than the linked arms of two people could span, and its crown spread out above all the other trees. It is hard to imagine that the pastures that stretch to the horizon were once filled with trees like this one.

And the possibility that such forests will ever return to the Bluegrass is fading quickly. As of 1999, beech bark disease had already reached West Virginia. It is only a few hundred miles from the folded hills of the western Appalachians to the heart of the beech's former range in Ohio and Kentucky. Once the disease is established here, there will be little chance that beech would be able to recolonize Kentucky's pastures and fields even if they were let go to forest. Wherever it strikes, beech bark disease reduces its victim to relict status—nominally present, but functionally invisible. The places where beeches remain healthy number fewer every year.

I stumbled unexpectedly across such a place one day late in 1999 on a hike with my friend Hanah on the hill behind her house in central Massachusetts. It was one of the rare uninfected stands that remain like islands in places where the disease has already run its course. Such places inspire an odd mixture of hope and fear—hope that they might be immune and fear that somehow they simply had been missed. In either case, I was glad to find them.

The sun was already low in the sky when Hanah and I emerged suddenly from a canopy of oaks into the beech forest. The change in the quality of the light as we moved into beeches was almost startling. Under the oaks the November air had been flat and cold, but the smooth shining trunks of the beeches seemed to give the air a warm glow. The leaves that still clung to the lower branches fluttered like small golden flags in the afternoon sun. We climbed faster and faster through almost nothing but beeches to the crest of the ridge and collapsed into a thick carpet of fallen leaves. All around were tall grey trunks and bright golden leaves, with a scattering of deep green mountain laurel underneath. We leaned back and watched the sky move in streaks through the dense canopy of clean, gray branches. We stayed there for a long time, taking it all in and wondering how many more years we might find this kind of light here on this hill on late afternoons in November.

Even with the devastation the disease may soon bring to this, beeches are unlikely to become rare here—let alone go extinct. But when beech bark disease infects a stand of beech, things change. The smooth gray trunks that rise luminous and clean through healthy forest become camouflaged with bark made rough and ordinary by the disease. Beeches under stress probably produce a much smaller nut crop than they once did, and many animals feel the loss of the food that healthy beeches provide. As the disease progresses, thinning crowns and dead trees let other species—sometimes invasive aliens—become established in the forest. The impacts of these changes are real, but are often so subtle as to escape the attention of all but the most observant. The forest has a way of obscuring wounds of this kind, leaving few scars that are visible to the untrained eye.

Though beech bark disease and hemlock woolly adelgid are alone causing fundamental changes in the character and composition of the landscape, the full impact of nonnative pests and diseases on the eastern temperate forests of North America goes far beyond these two invaders. The changes sweeping through these forests reach back nearly a century to a catastrophe that most of us know only through inference and history books. Only this kind of perspective can make it clear that the introduction of one pest after another is not a series of more or less manageable changes. It is a single century-long avalanche of degradation.

## Ghosts in the Forest

Though beech and hemlock are still abundant in many places throughout the eastern forests, there are a few trees native to this part of the world that live only in the eyes of those who know how to look for ghosts. First among them is the American chestnut, a tree that once accounted for a quarter of the standing timber in the forests of the eastern United States. It was decimated in the first half of this century by the chestnut blight, a disease caused by an introduced Asian fungus. Though chestnut persists in a few isolated groves planted beyond its native range and in the still-living roots and sprouts of victims of the blight, the tree that was once more abundant than any other is functionally extinct and virtually invisible.

Given that eastern and Carolina hemlock and American beech may soon be similarly reduced to shadows—present but without any real ecological weight—there is much to learn from the American chestnut. What role did it play in the human and the ecological economy of the region before the blight? How did society respond to the threatened loss of such an important species? And how do we see its loss now, fifty years after the destruction of the last chestnut forests?

The American chestnut was in economic terms the redwood of the East. Fast-growing, rot-resistant, and easily worked, it was the tree on which a significant portion of the country's early wealth was built. Early settlers made their homes and fenced their pastures with chestnut's durable wood and gathered its prodigious annual nut crop both for their own use and to sell in the city markets. The tree was a steady source of income for many farming families in a cash-poor rural economy.

While its wood and its nuts were what made chestnut valuable to the first arrivals in unsettled land, its resilience was what kept it prominent even once all the old growth in an area had been removed. The tree's ability to resprout vigorously from the roots meant that unlike most other timber species, chestnut increased in abundance in response to repeated logging. By the mid-1800s, chestnut had become the most important commercial timber tree in the East and was used for everything from railroad ties to fine furniture. At the end of the nineteenth century, chestnut was so abundant as to constitute half the standing timber in the state of Connecticut and up to 80 percent of the stems in some parts of the core of its range. It was a mainstay of the lumber industry in an era when wood was the primary building material of the American economy.

The ecological role of chestnut, however, is not as clear as its economic role. As one researcher has noted, "Scientific studies of forest community structure were undertaken only occasionally before the decline of the chestnut, [and] . . . only scattered references exist in the literature documenting chestnut as a wildlife food plant." Even so, it is probably safe to assume that its consistently large annual nut crop was an important food source for animals such as turkey, bear, and deer, which relied on the nut production of oaks, hickories, beeches, and chestnuts. Chestnut also was an important understory species as well as a canopy dominant, growing slowly in the shade of larger trees until a gap opened up in the overstory. This meant that chestnut often filled gaps in the forest more quickly and effectively than species that grew only from seed, particularly in areas where aggressive brushy species such as mountain laurel were likely to shade out young seedlings in forest openings. Even in the absence of comprehensive data about chestnut's role in pre-blight ecosystem function, the available evidence clearly indicates it was an important component of forests across its range.

Reports of a mysterious chestnut bark disease first came from the New York Zoological Park in 1904. A few years earlier, tree breeders at the park had imported several species of Asian chestnut (with larger nuts than the American chestnut, but a much smaller growth form) in the hope of producing a hybrid that offered both timber and large nuts. Unfortunately, the Asian trees brought a blight fungus with them. This pathogen, which did no significant harm to its coevolved Asian hosts, proved highly damaging to the related American species that had never before been exposed to the disease. The American chestnuts in the park were dead within a few years, and the blight spread with the winds.

By 1911, trees were dying from Massachusetts to Pennsylvania. No one had found any means of stopping the disease, or even of protecting indivd- ual trees, and it had become clear that the blight had the potential to destroy the chestnut across its entire range. In recognition of the tremendous economic value of the species—the governor of Pennsylvania at the time calculated the worth of the standing stock of chestnut in his state alone at $40 million—agricultural officials in states with significant stands of chestnut went to heroic lengths to combat the disease.

In 1911, the Pennsylvania State Legislature passed a bill appointing a commission to "ascertain the exact extent of the blight, and to devise ways and means through which it might, if possible, be stamped out." The legislature appropriated $275,000 (a significant sum at the time) to fund the effort. The commission conducted field surveys from June 1911 to early in 1912 and then called a conference to discuss their findings and investigate possible remedies. Representatives from all the affected states were invited, and the governor of Pennsylvania himself opened and led the proceedings.

The commission laid out its plan: It would determine the westward extent of the infestation, cut a break in the forest a mile wide to stop its advance, and patrol the uninfected zone vigorously for spot outbreaks, while simultaneously pursuing every possible avenue of research into controlling the disease. The commission was granted authority to enter and destroy trees on private as well as public land and had broad powers to design and enforce quarantine programs for chestnut nursery stock. It had the cooperation of other state governments and the support of the best minds in forestry science and plant pathology. The members of the commission confronted the blight with all of the considerable resources placed at their disposal. And they lost.

The blight continued its spread south and west, advancing an average of nearly 30 miles a year. By 1950, it had overrun even the farthest reaches of chestnut's native range, killing all but the roots of nearly every chestnut in the land. In barely half a century, this disease had reduced to little more than recurrent sprouts a tree that had once shaped the fundamental character of the entire eastern temperate forest. The American chestnut had become a ghost.

There are those, however, who believe in ghosts. The people at the American Chestnut Foundation are convinced that this particular ghost may one day be brought back to life. They and others are using a variety of techniques—from backcrossing the American chestnut with its resistant Chinese cousin to propagating other strains of the pathogen that render the blight less virulent—to create a chestnut that is able to survive even in the presence of the blight. They may be nearing success. If the genetics of disease resistance prove amenable to the gentle persuasion of selective breeding or the fire-with-fire tactics of virulence-reducing pathogens, the American chestnut could soon be granted a second chance.

But no matter what the outcome of this fifty-year resuscitation effort, the forests of eastern North America have been irrevocably altered. Even if the American chestnut is brought back as a species, it will probably never again dominate the landscape across its range as it once did. The oaks and hickories and other species that took over after chestnut was destroyed have created a new ecosystem, and a chestnut probably won't be able to reassert dominance within it. The chestnut ecosystem is gone, and no amount of breeding will bring it back.

And even if chestnut were restored to some portion of its former range—and if beech were saved and hemlock rescued, and all the other species under assault were somehow granted a second chance—the forests of the East would continue their decline so long as new pests kept being introduced. Especially given the fact that many forest pests are difficult or impossible to control, *prevention* must be the keystone of forest protection. But while keep-

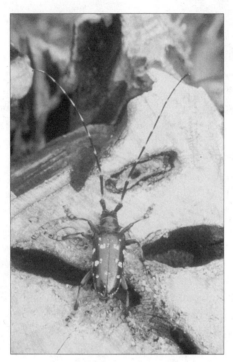

Asian longhorned beetles kill maples and other trees by riddling their trunks with holes and have the potential to devastate the hardwood-dominated forests of eastern North America. (USDA APHIS)

ing out the next major invader is the kind of goal that everyone ought to be able to agree on, the record shows that it has never been that easy.

## Free Trade and Forests

Ingram Carter was worried about the maple trees on his block. Their trunks were full of holes, and there were inch-long beetles crawling all over them. The trees didn't seem to be doing all that well overall—many branches were dying, and the leaves had a thin, anemic look to them. But it wasn't until a few months later that word got to the federal Animal and Plant Health Inspection Service (APHIS) that an unknown pest was killing trees in this Brooklyn neighborhood. In August 1996, APHIS sent a team to investigate.

What they found was truly cause for alarm. The insects were a species of Asian longhorned beetle, a group of organisms never before found loose in North America but known to cause significant damage to a variety of trees in its native China by eating out their heartwood. From what was happening in Brooklyn, it seemed that they thrived on North American trees as well.

Though APHIS is the federal agency with responsibility for keeping new pests out of the country, it was caught entirely off guard by the infestation and spent the next few months trying to determine an appropriate response. In late October, it convened a Science Advisory Panel to evaluate the implications of the invasion and recommend a course of action. The panel was unanimous in its conclusion: Impose a strict quarantine in the infested areas and launch an aggressive search-and-destroy mission immediately. Shortly thereafter, crews of city, state, and federal workers began combing both the Brooklyn neighborhood where the beetles had first been found and a second location on Long Island that had been reported shortly thereafter. Entire neighborhoods were stripped of trees as crews marked, cut, chipped, and burned hundreds of infested maples, horse chestnuts, elms, and other species.

But even more quickly than the crews were able to clean up known infestations, new reports kept coming in. In 1998, a series of infestations were discovered just outside of Chicago, and crews began taking down trees in several area neighborhoods. In New York, it quickly became apparent that the scale of the problem might be much larger than originally thought. Consequently, APHIS brought in professional tree climbers from a smokejumper crew in California, and by the end of 1999 they were climbing two thousand trees a month looking for new infestations. Unfortunately, they found what they were looking for. Another two sites were reported in Queens, New York, in early 1999. And in August of that same year, two dozen infested trees were discovered in Manhattan—only half a mile east of Central Park.

At first no one was sure how the beetles had gotten into the country. But after some investigation, APHIS discovered that many of the infestations were directly linked to solid wood packing materials used to ship goods from China. One of the Chicago infestations, for instance, was traced to a small manufacturing company that imported parts from China. Beetle exit holes were found in discarded crates behind the company's headquarters. The beetle larvae tunnel into the heartwood of the tree species they attack, and pallets and crates made from infested trees were a perfect vector for transporting these pests to North America. The problem is, when the larvae are inside the wood, they are very difficult for APHIS inspectors to detect.

For many years, APHIS had required only that solid wood packing be free of bark and obvious rot. In response to the Asian longhorned beetle crisis, it pushed through an emergency change in regulations requiring that all solid wood packing from China be treated with heat or chemicals in order to kill all internal pests. But they should have closed this door twenty years earlier. According to a story aired on CNN in late 1999, an entomologist working for the U.S. Department of Agriculture (USDA) warned APHIS in 1979 that many dangerous pests could easily be brought into the country in untreated wood. In fact, APHIS found Asian longhorned beetles in solid wood packing material from China in 1990—but inspectors simply fumigated the cargo and let it go. By the time the infestations were discovered in Brooklyn and Chicago, the beetles had probably already been in the country for five or ten years.

By the end of 1999, federal, state, and local authorities had spent nearly $10 million on eradication efforts. Clearly, much more was needed, for there seemed to be no end in sight. Reports of beetles were coming in from warehouses all over the country, and the probability that there were unknown infestations elsewhere in the country was very high. Back in 1996, a truck loaded with beetle-infested firewood from Brooklyn had been intercepted—at the Canadian border. No one knew how many truckloads of infested wood had already been missed. The situation was quickly becoming critical.

What was missing was a genuine commitment on the part of the USDA (APHIS's parent agency) to eradicate the beetle at all costs. According to Dr. Faith Thompson Campbell, an invasive species specialist with the American Lands Alliance, APHIS has mounted a response that does not even begin to address the magnitude of the crisis. Rather than declaring a large-scale quarantine, tracking down and inspecting all areas to which Chinese goods have been shipped, and pouring all available resources into eradication—as it did for many years, for example, for repeated invasions of agricultural pests like the medfly—APHIS largely limited its efforts to scouting for beetles in the vicinity of known infestations. Its approach is entirely reactive, and unless it starts to take a proactive approach, it is likely to ride the tail end of the infestation from now until whenever it becomes completely unmanageable. Escape into the wild might already be a foregone conclusion.

The implications of escape are staggering. The likely cost of tree replacement in New York City alone would run into the hundreds of millions of dollars, for maples are some of the most common shade trees in this and most other American cities. And as trees were cut and replaced with saplings of other species, an entire generation of city kids would grow up without trees.

But the potential loss to cities pales beside that to forests. One and a half *billion* forest trees in New England alone are at risk. Another 800 million are threatened in New York. Billions more could die across Pennsylvania, Ohio, Indiana, Michigan, Wisconsin, and Minnesota. White birch. Yellow birch. Black birch. American beech. American elm. Slippery elm. White ash.

Green ash. Many of the western hardwoods are susceptible as well. Bigleaf maple. All the cottonwoods. Quaking aspen. Bigtooth aspen. There really wouldn't be much left.

Not much, that is, but the invasive alien species that are already in the understory and at the margins of many of the native forests of the northeastern United States. Increasingly, weed trees like tree-of-heaven and shrubs like Japanese barberry and European buckthorn are what tend to move in when a gap opens in the forest. Since many of the native trees that are the most at risk from Asian longhorn beetle are the same species that filled in the gaps left by the other pest-driven native tree diebacks, there aren't many vigorous native tree species remaining that could step in should the beetle create large gaps in the forest. With the arrival of the Asian longhorn beetle, the cumulative impacts of a century of forest pests might well end up destroying the forest itself.

This was the first chapter of this book to be started, and the last to be finished. I put it off, I think, because I could not face the consequences of the conclusions I knew I would have to come to if I were to finish the chapter honestly. I did not know how to write about hemlock woolly adelgid and

Sugar maples throughout eastern North America are at risk from the invasion of the Asian longhorned beetle. (Gretchen Smith)

beech bark disease and Asian longhorned beetle in a way that was more than just words on paper. I didn't know how to write about the fact that my grand-children may never drink sap from the galvanized buckets that today hang from maple trees sometime in the first week of March. I didn't know if I could describe faithfully the luminous gray and translucent green of a beech forest in May. I didn't think I could find the words for the cool, lacy light under a hemlock grove in August. And I knew that I would never be able to convey how much I would miss the slow orange fire of the maple-covered hills in late October.

I did not know how to write about the death of the forests where I grew up. This is my home. How can you put something like this into words?

CHAPTER 7

# Guilty until Proven Innocent
## *Preventing Nonnative Species Invasions*

The world is going global. International commerce and travel are growing exponentially, and trade is destroying natural barriers between ecosystems just as fast as it removes barriers to the movement of goods. In a world whose motto seems to have become "No limits," the consequences of this loss of ecological isolation are seldom given much thought.

But in one place after another—across the United States and around the world—an alarmingly predictable pattern has emerged. As global trade increases, more and more goods and people move in and out of a region. Periodically, someone notices a new pest or disease that is killing native trees or clogging up rivers or obliterating native grasslands. More often than rot, the problem is recognized after it is too late to get rid of the new invader, so it is ignored and the damage increases. Against the worst of these invaders, conservation organizations or government agencies mount various often-uncoordinated suppression efforts on a few scattered fronts. In a few cases, biologists and land managers achieve a degree of long-term peace through biological control and better land use. But much of the time, control efforts fail to fully suppress the pest, and invaders become permanent residents, adding to an accumulating burden on local ecological integrity and native diversity. And in the meantime, a dozen more invasive alien species have probably been introduced.

Given the current lack of awareness about the threat of biotic invasions from careless trade, increasing globalization will inevitably bring more invasive pests. Unless human-made substitutes are created for the natural ecological barriers that are falling to globalization, the growth curves of trade and biotic invasions will rise in tandem. Today's existing invasive species problems are likely to be dwarfed by what another half century of careless ecosystem mixing may yet bring. Even in the United States, the world's largest, strongest economy, invasive species problems already have out-stripped the resources available to deal with them. Economically as well as ecologically, the United States—and the world—cannot afford an ever increasing flood of new invaders.

Therefore, prevention should always be the first choice for dealing with invasive species problems. Without prevention, all other solutions are inad-equate. Some damaging invaders start as intentional introductions that have unintended consequences; others are hitchhikers, moving unnoticed with cargo or with other species. These contrasting modes of introduction require different approaches to prevention, and a national comprehensive strategy to prevent harmful bioinvasions must address both types of species move-ment. More effective prevention methods have to become an integral part of how society works while there are still invasions to prevent.

## Thinking about Risk in a Biological Context

Most countries deal with public risk by prohibiting or regulating anything that is considered excessively dangerous behavior. In the context of pre-venting species invasions, prohibition is currently limited essentially to her-bivorous insects, plant pathogens, and diseases of vertebrates. Federal law prohibits importations that are deemed likely to bring in these pests. There are several problems, though, with the way most people define and evaluate risk in a biological context. The most fundamental issue has to do with a misunderstanding of the nature of risk itself as it applies to importation of nonnative species. Because segregating the good from the bad in importa-tions is often difficult or costly, society accepts the risk that prevention sys-tems will not work one hundred percent of the time. However, the nature of biological invasions is such that this approach is often far short of adequate.

For example, there are many highly dangerous pests of trees that could move between Siberia and North America on unprocessed logs. It is techni-cally impossible to ensure that there are not egg masses of pest moths stuck inconspicuously on a few logs in boats carrying thousands of logs. Similarly, it is impossible to ensure that the logs are free of internal pests such as lar-vae of bark beetles or borers. If such shipments were to be made only once, intensive inspection and some luck might be enough to prevent the impor-tation of a damaging invader, for the odds are good that a single introduc-tion will not result in establishment. However, when such shipments become frequent—occurring hundreds or even thousands of times every year—the possibility that one or more boatloads will be the source of a major new invasive pest essentially becomes a certainty.

This follows from simple probability. If, for example, there is a 1 percent chance that a given log will have damaging live insects associated with it, and if there is a one percent chance that an inspector will fail to find the pests, and if there is a 1 percent chance that such pests will establish and become invasive—all conservative estimates of the risks of these events—then there is one chance in a million that a given log will be the source of a new pest. Is that level of risk low enough? Suppose that three ships move logs each day from Siberia to Oregon and that each ship carries 10,000 logs. Within ten years an invasion is essentially a statistical certainty, as in that time nearly a million logs would have been moved. Given that such invasions generally cannot be stopped once started, and given that the forests of the Pacific Northwest support multibillion-dollar industries and are of incalculable ecological value, even one chance in a million may be too high a risk. What appears from an individual importer's perspective to be a vanishingly small chance is from a societal perspective a dangerous inevitability.

Compounding the issue of social versus individual risk is the fundamentally different nature of the consequences of making mistakes with biological introductions as opposed to other kinds of environmental risks. In the context of chemical pollution, risk abatement is based on the fundamental principle that risk of damage declines as exposure declines. For example, if altered production methods reduce the amount of toxic or carcinogenic materials that workers are exposed to by 95 percent, then the risk of illness for any given worker should be negligible. Threats such as pesticide residues on food and ozone depletion in the atmosphere have taught us to think about risk in terms of cumulative impact: If you wash your vegetables whenever you can and wear sunscreen more often than not, chances are good you will avoid the various cancers and other maladies that total negligence would invite. In other words, 99 percent is generally close enough to perfect, and the 1 percent you get wrong usually doesn't cause any appreciable harm.

But the permanence and the self-spreading nature of biological invasions means that if a catastrophic invasion occurs, it doesn't matter if the probability of such an event was one in a million or one in three. In a biological context, a 95 percent reduction in risk doesn't mean that 95 percent less harm will occur; it simply means that there is now a 5 percent chance that 100 percent harm will occur. Invasions are put off, but when they happen, their effects are unmitigated. If the importation of a risky cargo is repeated many times, it is almost inevitable that an invader will eventually be introduced. Thinking in terms of risk in the context of species introductions should not be so much about minimizing risk in all situations as about avoiding high-risk situations at all times.

A similar model of risk analysis applies to intentional introductions of nonnative species. Just as quarantine and inspection protocols can reduce but not eliminate the possibility that stowaway species will accidentally be introduced and become pests, evaluation and screening of proposed species importations will never be entirely free of error. Given enough species introductions, even the most conservative importation program will eventually allow the importation of seemingly harmless species that turn out to be highly damaging. The goal of prevention efforts cannot be to eliminate all risk, but a clear understanding of the nature of risk in a biological context can help to minimize the frequency with which the system fails.

## Preventing Accidental Introductions

In theory, preventing accidental introduction of pests is something everyone can agree on. There is a general social consensus that certain kinds organisms—mostly plant and animal pathogens, insect pests, and obvious weeds—are not welcome. Where disagreement begins is on how high to raise the bar. How much benefit is society willing to forgo and how much regulation will people put up with in order to avoid accidental pest introductions? Thoughtful evaluation of these choices demands a clear understanding of what constitutes high- and low-risk activities, a realistic assessment of the power of inspection systems to erect protective barriers, and a creative approach to developing proactive programs to identify and reduce risk.

### High- and Low-Risk Importation Patterns

Importation of raw logs from Siberia to western North America is inherently risky in that ecological similarities between the two regions make it highly likely that many pests that do arrive on imported logs will thrive at the expense of native tree species. By contrast, other kinds of importations are by nature less likely to result in the introduction of a pest species. For example, goods moved from tropical countries to temperate regions with freezing winter temperatures are safer. Few if any species from a tropical climate could survive the winter in a temperate zone. Mahogany logs imported to Oregon, for instance, would pose little threat to the local forest ecosystem, even if they occasionally brought their bark beetles or borers with them. These species would have no local hosts and would be unable to survive.

In general terms, the relative degree of risk of a given pattern of importation depends on the ecological relationship between the source and the
destination of the goods moved, as well as on the volume and the nature of
the cargo itself. Importation patterns that pose a high risk of pest introduction involve commodities that are moved between regions with similar climates and plants, commodities that are moved in large quantities on a regular basis, and commodities that readily conceal pests. Conversely, the risk
of accidental pest introduction is relatively lower for commodities that are
imported from regions with less rigorous climates, commodities that are
moved to areas where similar plants are not present to serve as hosts, and
commodities that are moved in small quantities. Any estimate of the degree
of risk of a given action is based on an educated but ultimately subjective
assessment of the net effects of all the above factors.

Government efforts to reduce risks often entail placing conditions on
importation that regulate how and where different kinds of imported goods
from different points of origin must be handled. Importations that pose a
high risk of accidental introduction are generally regulated more heavily
than those that are considered low risk, but even with importation patterns
that are essentially guaranteed to carry pests, regulation is much more commonly used than outright prohibition. For instance, one common stipulation
of an importation permit is that a given item may be imported to certain
parts of the country but not to others. However, because 95 percent compli-
ance is not enough to prevent the introduction of new pest species, relying
on such limited permissions is a poor strategy for controlling risks of species
invasions. Ultimately, the only reasonable way to handle high-risk importations is to prohibit them entirely.

### Inspections as a Tool for Detecting Invaders

To slow the rate of uninvited nonnative species introductions, countries can
inspect incoming materials in an effort to detect stowaway organisms.
Inspectors at international ports of entry routinely conduct such inspections.
This usually consists of asking travelers if they are carrying such high-risk
items as fruit, seeds, eggs, or live plants, items that generally are either possible pests themselves or potential hosts of insects, or plant or animal
pathogens. Inspectors are also responsible for reviewing the documentation
and checking the contents of commercial shipments of goods, be they
tankers full of wood chips or a planeload of roses. This model of inspection
is largely aimed at slowing the spread of agricultural pests.

Given the large and increasing volume of food, ornamental plants,
wooden packing materials, and nursery stock shipped internationally every

year, inspectors typically are limited to spot checks of a few shipments. Inspections are therefore filters—not barriers—and they often miss nonnative invaders at either of two levels. First, the fact that only a tiny fraction of shipments (perhaps 1 percent) are actually inspected provides ample opportunity for organisms to be missed by simple noninspection. Although inspectors focus their efforts on high-risk shipments (making actual inspection efficiency somewhat greater than 1 percent), much is missed.

Second, even for shipments actually checked by an inspector, small or concealed organisms can easily be overlooked, especially if they are hidden—wood-boring insects in packing material, for example—or invisible, such as pathogens that have not yet caused visible symptoms. The chestnut blight, for example, entered the country on apparently healthy Asian chestnut seedlings. Even if these trees had been inspected, detection would have been impossible without the help of skilled plant pathologists (who are unlikely to hold jobs as port inspectors) and enough time to take and culture samples before shipments are released. Similarly, the invasion of the Asian longhorned beetle that is now killing maples in the northeastern United States was caused by the movement of beetle larvae inside boards in crates used to ship manufactured goods. Such internal organisms are extremely difficult to spot, even with concerted effort.

Inspection is effective only if the nature and volume of the cargo that inspectors are charged to examine are tempered by sensible policies regarding what can and cannot be brought into the country. Relying on inspection as a primary barrier instead of as a final check is akin to trying to filter a river of muddy water with a handkerchief. It will fail.

## Active Prevention of Species Movement

For species known to be dangerous, or pathways very likely to promote movement of unwanted species, governments can create proactive programs to reduce the rate of introduction. Some states have their own proactive efforts against particularly threatening species. For example, Hawai'i fears the brown tree snake will drive many of its remaining native bird species to extinction and has mobilized a major effort to prevent its invasion (see Chapter 13). These snakes have already exterminated most forest bird species on Guam, the site of the snake's initial invasion of the North Pacific Islands region.

Two approaches contribute to reducing the risk of moving snakes from Guam to new islands: snake-proofing the airports on Guam and inspecting cargo from Guam for snakes as it arrives on other islands. The fewer snakes there are at airports and in warehouses on Guam, the less likely it is that some will be mixed in with crates or other goods leaving Guam. And in Hawai'i, a combination of fences, snake-detecting dog teams, and protocols for responding to snake sightings both reduce the risk of transport and provide a means of detecting snakes should they arrive on a flight from Guam. In the longer run, though, the overall density of snakes in Guam must be reduced—by introduction of pathogens or other means—so that Guam's threat as a source of infestation to the rest of the northern Pacific islands is permanently lowered. Anything else is simply a holding action.

Another form of active prevention has been applied to the movement of aquatic species in the ballast water of cargo ships. The bilges of large ships plying the oceans are floating aquaria, picking up species (mostly marine invertebrates and plankton) in the coastal waters of one continent and then discharging them elsewhere when ballast water is pumped out at the port of destination. This route of entry became a concern in North America in the early 1990s following the discovery that zebra mussels had been accidentally introduced into the Great Lakes via ballast water. The high cost of the zebra mussel invasion in North America prompted the passage of a law to reduce the risk of such introductions by requiring ships to exchange ballast water in the high seas or treat ballast water with chemicals. These and other regulations, while not yet robust, have the potential to change what was a high-risk pattern of international trade into a relatively low-risk one, and such rules are some of the most powerful tools available to prevent nonnative aquatic species invasions. The extension of such practices worldwide is urgently needed.

## Invited Species That Become Invasive

Though the techniques for prevention differ, the same fundamental logic that applies to reducing the impacts of accidental introductions also applies to minimizing harm from deliberately imported species that become invasive. Keeping out accidental stowaways focuses on the composition and size of the regulatory wall erected; evaluating species for deliberate importation is concerned with when it is acceptable to open the gates. Unfortunately, the same kind of misunderstanding of the nature of biological risk that allows for one accidental introduction after another also prevents the creation of a comprehensive system of evaluation for proposed introductions. Until there is a fundamental shift in public understanding of risk that allows for a much more vigorous and proactive approach to both wall-building and gatekeeping, prevention efforts will only delay invasions, not prevent them.

There are two approaches commonly in use for the evaluation of proposed introductions. "Dirtylists" identify species that are known or presumed to be harmful and prohibit their importation and release. This approach to prevention is grounded in the assumption that species proposed for importation are innocent until proven guilty and is the principal approach used in the United States. "Cleanlisting" compiles lists of species for which the preponderance of evidence indicates relatively low risk of invasiveness or harm and allows for their importation without review. Australia and New Zealand have pioneered this approach to prevention. The track record of use of this second approach is too short to draw conclusions yet about its practical effectiveness. However, the fact that it substitutes a presumption of guilt for the presumption of innocence that underlies the dirtylist approach is very promising.

## Dirtylists and Their Shortcomings

Whether a dirtylist is at all useful depends largely on the length of the list and the thoroughness of the efforts to determine if it is reasonably complete. At its simplest, a dirtylist is merely a roster of a country's past mistakes—that is, of introductions already made and therefore largely irrelevant to prevention. If species are prohibited only after they have invaded and caused damage, there may be few opportunities to benefit from such lists by excluding the listed species from not-yet-invaded areas.

A more useful approach would be one in which the listing agency actively seeks information about species in advance of importation requests for the purpose of better recognizing species they should be dirtylisted, perhaps because they are pests elsewhere or have features that suggest they might readily become pests. Australia is presently trying to use proactive review of South African plants to dirtylist species likely to be invasive in Australia before they are requested to be imported. Information on whether the species is considered a weed in the native region and whether it occupies a wide range of habitats or reaches high densities can be useful in creating such protective dirtylists.

In the United States, the Animal and Plant Health Inspection Service (APHIS) has the statutory authority to review and prohibit the introduction of species deemed to be dangerous. The dirtylisting of the seaweed *Caulerpa taxifolia* in the late 1990s is an example of such an action. This plant was used in public aquaria in Europe until it was discovered to be highly invasive in the Mediterranean Sea. The threat of this species to United States coastal ecosystems was so obvious that APHIS acted to prohibit its sale in the United States. Many equally obvious threats, however, have not been addressed, for the agency's funding and political mandate are inadequate to take such action on a broader scale.

In a regulatory climate dominated by economics, any strongly proactive dirtylisting attempts are likely to run into political difficulties. Importers whose income would be affected if the range of importable species were reduced are likely to object vigorously to any such attempt to dirtylist anything but obviously damaging species. They are likely to argue, often persuasively, that they are suffering undue economic harm and that there is no clear proof that any particular suspicious species is risky enough to merit exclusion. Within a standard framework of risk analysis, the importer's arguments are sound because the risk that any particular species will actually become invasive is quite small. Unless there is nearly incontrovertible evidence of probable damage, few risky species will be dirtylisted based on this approach to risk evaluation.

Herein lies the fundamental limitation of dirtylists. Excluded species are the special category, not the default classification. This means that people have a right to bring in any new species they want, unless its importation has been proven so obviously detrimental that it warrants prohibition. The presumption of innocence and the focus on individual rights rather than collective risk define the dirty list approach and place such large constraints on its use as to render it of limited value. It is the direct equivalent of how pesticides were regulated before passage in 1947 of the original pesticide control act (FIFRA), which shifted the burden of proof from government to industry. Before 1947, government health officials had to prove on a case-by-case basis that a particular pesticide residue was dangerous. After the passage of FIFRA, pesticide companies wishing to market a pesticide had to prove the product was safe. Though the system certainly has its defects, the orientation it takes at least allows for the possibility of improvement.

## Guilty until Proven Innocent

If one takes the view that people do not have any intrinsic right to import nonnative species, then the burden should be on the importer to show that the species to be brought in will not cause harm. Since the ability to identify invasive species in advance is limited (see Chapter 5) and the consequences of being wrong are great, such a conservative approach—one that makes prohibition the default and permission the exception—is the only sensible strategy. This shifting of the burden of proof is the key to effective prevention.

Presumption of guilt is already the model for a few categories of species.

For example, in order to protect agricultural interests, the federal government prohibits the deliberate importation of all herbivorous insects and plant pathogens without specific authorization. Anyone who wants to bring in, say, a leaf-eating beetle as a specialized natural enemy of an introduced weed has to show that the proposed species importation will not harm agricultural crops or native species before permission will be granted to import and release the insect. This system is imperfect, but it provides significant protection by assuming that preventing damage to valuable agricultural and native species is at least as important as gaining the benefits that importing the new species might offer.

For importations that have no direct bearing on agriculture, native ecosystems are not given such protection. Virtually any species that is not likely to attack plants can be imported. A rancher could buy a pair of elephants from a game farm in South Africa and release them on his land in Texas (provided they are healthy, disease-free elephants). A fancier of nonnative pets could import every known species of tarantula and release them all in her backyard. Since neither large mammals nor carnivorous arthropods are specifically forbidden, the default status under which they fall is that of unregulated species. If laws existed to protect native species and ecosystems, such importations would not be allowed unless the importer could show that they would do no significant harm. A presumption of guilt for all proposed importations would be a significant step toward preventing the establishment of new invasive nonnative species.

## Cleanlists: A Proactive Approach

Such a presumption of possible harm is the principle underlying the cleanlist approach. All nonnative species are assumed to be potentially damaging, and only those that appear unlikely to become invasive are allowed to be imported. But even use of cleanlists may present difficulties. For small groups of relatively well known species like birds, mammals, and other vertebrates, cleanlisting is probably technically feasible. For large groups like insects and plants, it is not. There are so many species in each of these groups (many of them unknown) that to even make a list of all species—let alone gauge the potential invasiveness of each—simply is not possible.

However, the power of the cleanlist approach is not so much in the lists of permitted species that it generates as in the fact that anything not yet reviewed is presumed to be unsafe. The practical consequence of this approach would be to eliminate frivolous species importations, for only those with real social, economic, or ecological benefit would be worth the time and expense required to demonstrate noninvasiveness. Introductions would still happen, but only after careful consideration.

## Current U.S. Policy Regarding Nonnative Species

The inability of the U.S. government to formulate a coherent national policy on the importation of nonnative species (accidental or deliberate) derives largely from the fact that people have very different feelings about different kinds of organisms. As a consequence, the process of evaluation for each kind of species begins with a different set of assumptions about the relative threat a given class of organisms presents. These assumptions have their origins more in cultural history than in ecological science, but their influence over policy is profound. The United States' agricultural roots condition people to see most plants (especially beautiful or edible ones) as beneficial and most insects that eat plants as suspicious at best. It also predisposes us to look favorably upon most mammals, birds, and fish, especially game species. Most everything else simply fails to register. But invasive species come in many varieties, and effective prevention demands a level-headed and ecologically based approach to evaluating the potential benefit and harm of each proposed introduction.

### General Approaches for Various Kinds of Species

For the most part, nonnative plants are assumed to be innocent and beneficial—and therefore noninvasive. Some 25,000 species or varieties of plants are in commercial trade in the United States, and new species are brought in every year. Importation is forbidden only for a tiny handful of plants— under 200 species, most of which are crop weeds or parasitic plants. Nearly half the plants on the prohibited list are parasitic species, such as dodders or members of the genus *Prosopis* (woody plants related to mesquite), so the breadth even of this short dirtylist is much less than it seems.

Invasiveness in natural areas has long been poorly reflected in the species chosen for dirtylisting. However, the recent inclusion of several invaders of natural areas (such as the shrub *Mimosa pigra*, the Australian paperbark tree, and the invasive marine seaweed *Caulerpa taxifolia*) suggests the beginning of an expansion of the scope of prohibitions. Also, a few invasive waterweeds (such as some species of *Salvinia*, *Azolla*, and *Eichhornia*) and invasive grasses (species of *Pennisetum* like Kikuyu grass and African feather grass) have also been added to the list in recent years. Governmental policy toward potentially invasive plants is shifting, albeit very slowly.

As mentioned earlier, any species that attacks crop plants is viewed as a threat to an essential resource, and importation of any herbivorous insect or plant pathogen is forbidden unless specifically authorized. Plant importation is recognized as the major route of such invasions, and as a consequence, laws were passed early in the twentieth century in an effort to slow the accidental introductions of pest insects and pathogens that flooded into the country in the late nineteenth century when there were no controls on the shipment of plants. It is important to note, however, that inspections are only for insects and diseases; the plants themselves are not an object of scrutiny, provided they are not on the short dirtylist mentioned above.

Biological control is the deliberate importation of plant pathogens and herbivorous or carnivorous insects for control of weeds and pest insects. Agents proposed for control of pest plants are subject to intense scrutiny and testing under the plant protection acts, and each must be shown to do no significant harm to valuable species (cultivated or native). However, importation of agents for control of pest insects is not regulated, as existing U.S. laws address only the introduction of herbivorous insects and such potentially damaging indirect agents as hyperparasites (species that attack beneficial parasites and thus indirectly might promote injury to plants). Requests for permission to import biological control agents generally come from scientists who work for government agencies and public universities, and such importations are to varying degrees regulated by the federal government. Private firms also make requests to ship and sell parasites and predators for insect biological control in greenhouses and on high-value crops.

For many years, both the public and the government viewed importations of biological control agents for insect pests as inherently safe, wholly beneficial, and contributing to increased public safety by reducing the use of pesticides on crops. In more recent years, several cases of effects of biological control agents on nontarget species have been identified, and both the scientific community and the public at large have begun to scrutinize biocontrol introductions more carefully. A more nuanced assessment of the benefits and potential risks posed by such introductions is being phased in, but a need exists for a national biological control law, as other laws either do not provide specific legal authority or, like the National Environmental Policy Act's Environmental Impact Statement process, are a poor fit to a non-site-specific action like the introduction of a new species. A code of conduct on introduction of biological control agents has been developed at the international level for countries to consider when developing their national policies in this area and might serve as a useful starting place for such a law in the United States (see Chapter 11).

Importation of insect species for use as pets or for other commercial purposes is generally viewed favorably by the federal government. Butterfly houses import a variety of herbivorous butterflies because the public appreciates the beauty of such species and such endeavors can be profitable. Technically, such importations are not legal unless specifically authorized, because the species involved are all plant feeders and are therefore regulated by current plant protection laws. However, the fact that butterfly importations are routinely permitted highlights the nonecological foundation of importation policies—species that do not "look" threatening are more likely to be let in, regardless of their actual impacts. Whether butterfly houses will be a source of species invasions is not yet certain, as importation of butterflies for such purposes is a relatively new phenomenon. The likelihood is that in the long term, they will.

Pet arthropods, on the other hand, are often not plant feeders. Cockroaches and spiders are among the species sold in pet shops. If the species involved do not eat live plants, they are not regulated by federal plant protection acts in the United States. Some states, such as Florida, have recently become alarmed over the threat posed by such unregulated sales of arthropods and have passed state laws banning their sale, but the ease with which such species can be shipped across state lines significantly limits the effectiveness of regulation at anything but the federal level.

Importation of "wildlife" (roughly, terrestrial vertebrates) poses two problems, only one of which is addressed by current laws in the United States. First, nonnative wildlife can vector diseases of domestic livestock and native wildlife. As early as the mid-nineteenth century, veterinarians recognized the important role of imported animals as a means of bringing in new diseases and parasites, and the first U.S. wildlife importation laws (enacted in the late 1800s) aimed to reduce this risk by a mix of inspections, quarantines, and prohibitions. Quarantine and inspection were used with newly imported animals to determine their health and ensure their freedom from parasites. Prohibitions of importation of some species, or any species from some areas, were imposed to stop the spread of some key wildlife and domestic animal diseases such as hoof-and-mouth disease.

However, current laws do not address the threat of invasion of natural areas by imported wildlife species themselves. Invasive species can threaten the biodiversity of the invaded area by eating or competing with local species or by changing their habitat (see Chapter 5). Importers currently do not have to consider such impacts when petitioning for the right to bring a new species into the United States. Rather, they must show that the imported species is not a disease vector and that it is not a rare or threatened species in its area of origin. If these criteria are met, there is no federal barrier to prevent game ranchers from bringing in favored species for hunting, exhibition, or sale as pets. (Some states do prohibit this type of activity, but again, state laws are of limited effectiveness.) Wild elephants roaming the west Texas plains seems far-fetched, but wild emus in Georgia appear to be a reality following the collapse of the emu-rearing craze. If a supplier of pet lizards wanted to start farming all the lizard species of Asia on 60 acres in Arizona, current laws would not forbid it. The threat that such a business would pose to native lizards would be large and obvious, but would be perfectly legal under existing statutes.

A series of disasters in the nation's waters have heightened governmental awareness of the impacts of introduced species on aquatic environments and resources. Zebra mussel damage to the water pipes of commercial enterprises using surface water sources runs to the billions of dollars, and zebra mussels threaten dozens of species of native mussels with extinction. Other aquatic species threaten to degrade (or have already degraded) the productivity of shell and fin fisheries in Chesapeake Bay. Sea lampreys have devastated the lake trout fishery of the Great Lakes. Pathogens of oysters and shrimp potentially found in materials exchanged among aquaculture businesses could damage the wild stocks of these valuable species in the eastern and southern United States.

These problems have led recently to a call for new laws aimed at reducing the threat of introduction of aquatic nuisance species. One part of the problem—movement of pests in ship ballast water—has been partially addressed by a new law requiring ships entering the Great Lakes to treat their ballast water chemically or exchange ballast water on the high seas. The next logical step is to extend this requirement to all U.S. ports and eventually to all shipping by means of an international treaty. The intent of such legislation is to prevent the movement of coastal or inland species from one continent to another. Other aspects of the aquatic species threat, however, remain unaddressed.

Fish reared commercially (for food or for sale to hobbyists) may themselves become unwanted invaders of native habitats. Blue tilapia, for example, has invaded the habitat of threatened native fish in parts of the western United States. However, because fish farming is an agricultural enterprise, the species of fish under cultivation are considered a valuable resource and the threats they pose to native ecosystems are largely ignored.

State departments of fish and game have for over a century routinely transferred species outside of their historical ranges to diversify or increase angling opportunities for the public. Many of these introductions have decimated native fish populations. For example, the widespread introduction of rainbow and brown trout has depressed or eliminated populations of many local trout species, especially in the western part of the United States, an area rich in local native trout. Some states have begun to rethink such deliberate introductions in light of the harm that nonnative fish can do to local fish species.

However, fish rearing for the pet trade or aquaculture complicates matters by providing other routes of invasion of nonnative fish. Establishment in Florida of various African tilapia species in the 1960s and 1970s illustrates the problem. These species simultaneously created new fisheries of commercial value and damaged existing stocks of various native fish. The problem of nonnative fish introductions ultimately stems from the fact that agencies making importation decisions generally focus on the specific resources to be gained—the new game or food fish—rather than on the integrity of the ecosystem as a whole.

## The Legal Framework

Most current alien species control laws were developed to protect important natural and agricultural resources, not to prevent introduction of invasive species per se. Any protection of native diversity and natural ecosystems is incidental. Thus while the law in some areas is clear and has a long history of implementation, in other areas there are important threats that are being ignored.

Various animal quarantine laws, some of which date back to the late nineteenth century, protect domestic and wild vertebrates from the introductions of pathogens on imported stock via inspections, quarantines, or veterinary certificates.

The Lacey Act was passed in 1900 to prevent interstate commerce of animals killed illegally in any state and importation of certain types of wildlife determined to be injurious to agricultural and horticultural interests. It also provides foreign countries with assistance in enforcing their wildlife conservation acts. The act also prohibits the importation of a short dirtylist of undesired species, including mongooses, fruit bats, zebra mussels, and various other birds, mammals, reptiles, amphibians, fish, mollusks, and crustaceans that have been determined to be injurious to agriculture, horticulture, forestry, or wildlife resources, or to aquatic and terrestrial vegetation upon which wildlife depends.

The Plant Quarantine Act was passed in 1912 and gives the secretary of agriculture authority to regulate the importation or interstate shipment of nursery stock or other plants or plant parts when necessary to prevent the introduction of plant pests. It is important to note, though, that the act does not deal with whether the plant itself might be a pest, but only whether a particular shipment of plants is free of herbivorous insects and plant pathogens.

The Federal Seed Act has since 1939 regulated purity and truth-in-labeling of imported crop and vegetable seeds and set tolerances for contamination by nine species of crop and rangeland weeds. Note that the intent here is only to maintain product purity for the consumer and to limit levels of, but not exclude, a few important weeds.

The Organic Act was passed in 1944 to promote the eradication of incipient infestations of certain pests. This law provides APHIS with the authority to identify and act against new invaders. However, APHIS's ability to respond effectively under this statute is significantly constrained by both lack of funding and lack of political support.

The Federal Plant Pest Act of 1957 provides APHIS with authority to regulate the introduction of economically important parasitic plants. Note that this does not identify invasiveness of natural habitats as an issue over which regulatory authority is granted to the agency. The act also prohibits the importation of plant pests (basically, any herbivorous insect or pathogen), except as covered by a permit from the secretary of agriculture. Pests of all plants are covered, not just those affecting economically valuable plants. This law both prohibits the importation of potential pests of plants and also provides the authority for the government to regulate the importation of plant-feeding insects and pathogens as weed biological control agents.

The Federal Noxious Weed Act was passed in 1974 to permit the exclusion of designated foreign weeds that do not occur in the United States or have limited U.S. distributions, allow for the regulation of the interstate movement of species under quarantine, and promote cooperative eradication activities between federal, state, and local governments and eradication efforts on federal lands. Seed shipments are exempt from this act.

The Agricultural Quarantine Enforcement Act of 1989 prohibits the shipping by first-class mail of any plant, fruit, vegetable, or other matter quarantined by the Department of Agriculture. However, First Amendment rights require a search warrant to open packages for inspection, so enforcement is weak.

The Non-Indigenous Aquatic Nuisance Prevention and Control Act was passed in 1990 in response to the zebra mussel crisis and seeks to prevent invasions associated with international shipping. It requires that ships entering U.S. waters either treat their ballast water chemically or exchange it on the high seas. Unlike all other invasive species–related legislation, this act is focused not only on protecting particular economic resources from pests, but also on protecting native ecosystems from a specific pathway of invasion.

## Developing Better Policies for Prevention

Most laws in the United States that regulate the importation of nonnative species were drafted long ago and address a narrower set of issues than those of current concern. As mentioned earlier, these laws do little to address invasiveness of species that are not threats to agriculture or the health of livestock. In general, attempts to meet new concerns with old tools are often unsatisfactory, because old laws simply do not provide the authority needed to solve new problems or are incomplete or contradictory when applied to new concerns.

The development of a better system of prevention hinges on several key issues. Given the current international emphasis on reducing barriers to trade, perhaps the most pressing issue is how to balance the economic pressures for free trade against the need to restrict trade so as to protect biodiversity. Accountability is another crucial factor, and payment for mitigation of harm caused by damaging invasive species has to be linked to the businesses that profit from importing or selling nonnative species or products that transport damaging species. These two central issues—integrating protection measures into trade policy and creating a funding mechanism for mitigation of damage—serve as a platform on which sound prevention legislation can be developed.

### The Dilemma of Free Trade

Laws prohibiting the entry of agricultural products have at times used the potential threat of pest invasion (real or imagined) to achieve market protection. Under current world treaties on trade, such practices are subject to challenge by the country whose products are excluded. Japan has long restricted the importation of American apples because it claims they might vector pest insects not yet present in Japan. Washington State farmers deny this assertion and see Japan's action as market protection. American farmers recently pressed for permission to increase apple sales to Japan, and as of late 1999 the dispute was still unsettled. Conversely, the United States, fearing invasion by tree-killing Asian longhorned beetles, prohibited the use of packing crates made of untreated wood for shipping of commercial goods from China to America in 1998. China, fearing a loss of business, objected, and details of the settlement were still being negotiated in late 1999. Conflicts between trade and prevention of biotic invasions are becoming more and more common.

Efforts to limit trade to prevent pest invasion and protect local biodiversity will increasingly have to be defended in World Trade Organization dispute panels against opposing interests whose economic concerns would be hurt by such restrictions. How will the still-uncertain ecological and economic costs of potential invasions be judged against the well-defined economic costs if permission to import a specific good is denied? Part of the solution must come from a major change in public perception. Until a large number of people both recognize the link between increased trade and rising numbers of invasions and appreciate the ecological and economic consequences of invasive species, the problem will only get worse.

Just as important as public awareness, though, is a solid factual basis for claims of harm caused by invasions. The scientific literature provides ample evidence for the negative ecological impacts of invaders, but only a small segment of the population sees environmental degradation as a first-tier issue. What is missing is a broad body of knowledge of the social and economic consequences of nonnative species invasions. Since loss of the economic and social amenities that natural systems provide is often incremental, few people realize just how enormous a weight pests already present impose on economic use and cultural enjoyment of the natural world. A recent study by David Pimentel of Cornell University attempted to quantify the negative economic impacts of invaders and estimated that the invasive nonnative species currently cause about $137 billion in damage annually in the United States alone. Many more studies of this kind are needed, because until prevention speaks the language of economics and well as ecology, it will consistently take a backseat to free trade.

## A System of Accountability

When invasions happen—and they will continue to occur even under a well-designed prevention program—who should pay to mitigate their effects, which at times run to millions or even billions of dollars? Under current law, it would be difficult or impossible to assess blame and impose damages. Will it ever be possible to figure out exactly which ship's ballast water brought zebra mussel larvae into the Great Lakes? Are government agencies responsible for the impacts of government-sponsored introductions with unforeseen ecological consequences? Unfortunately, society operates on the assumption that unless specific violations of importation regulations occur, invasions are no one's fault. Even if it was willful negligence that led to the introduction of a new pest, the offending party is not responsible for anything more than the fines associated with such violations. A shipping company could be fined for failing to flush its bilges before entering coastal waters, but it could not be held responsible for the costs imposed on society by any new invaders that it happens to carry in its ballast water.

In any case, it is in most cases impossible to link most invasions to the persons or corporations responsible. One alternative would be to take a cue from the insurance industry and levy a tax on international shipments in proportion to their potential to cause biotic invasions. All groups benefitting from trade in nonnative species (or types of cargo that could carry them) would be required to pay into a mitigation fund based on the volume and nature of the cargo they carry. Money raised would then be held in reserve to allow rapid funding of efforts to eradicate new invaders or develop biological control programs for pests if they are only detected after eradication is no longer possible. Such a system would require that businesses and individuals trading in risky organisms take collective responsibility for resolving the problems that arise from their activities. Such a pooled system would sidestep the obvious defense made by such groups that only a small percentage of nonnative species become damaging and would instead recognize the even more compelling truth that these few species, while not numerous, affect large areas and impose huge economic and ecological costs.

## Placing "Filters" between Ecosystems and Trade

Even though outright prohibition of all vectors of species movement is not feasible, there are a few measures that would serve to reduce the number of new introductions per year. Proposed changes in federal laws or regulations regarding prevention of introductions might logically be organized around a series of "filters" designed to protect native ecosystems from damaging invasion. Since the legal context of species importations and trade regulation in general is constantly changing, the following points—or filters—provide a brief indication of what is needed without proposing specific, detailed pieces of legislation.

1. Mandate the preferential use of local native species from local seed sources (see Chapter 14). *Possible new legislation and/or regulations:* An executive order requiring that all federal agencies use only native species for landscaping, revegetation, and other activities, unless there is a compelling reason to use a nonnative species, and requiring the same of all state and local units of government and private organizations that receive federal funds.

2. Mandate the preferential use of nonnative species that, based on experience, appear to be highly unlikely to become invasive. *Possible new leg-*

*islation and/or regulations:* A law prohibiting the deliberate introduction of any new nonnative species without review and either providing federal funding for evaluation of requests to import species or mandating payments by importers (or both).

3. Develop robust inspection services to detect and exclude unwanted stowaway organisms. *Possible new legislation and/or regulations:* Revised regulations giving postal and customs inspectors the authority and the funding to inspect both domestic and international mail and cargo suspected of carrying prohibited species; new laws that impose substantial penalties for the interstate or international shipment of prohibited species without permission.

4. Identify high-risk species or vectors of introduction. *Possible new legislation and/or regulations:* Increased funding for APHIS, the federal agency charged with these matters; new legislation giving APHIS an express mandate to prohibit permanently and totally the importation of any species whose life history suggests the potential for invasiveness and to control high-risk pathways of introduction.

5. Create a remediation fund that would be available for rapid response to new invasions. *Possible new legislation and/or regulations:* A law imposing a tax on all economic activities that serve as vectors of species introduction; new regulations and interagency agreements that create a system of detection and response to new invasions using funding generated by the new tax.

## The Federal Executive Order on Invasive Species

The process of developing a series of protective filters between invasive species and native ecosystems has already begun. On February 3, 1999, President Bill Clinton took a significant step toward creating a comprehensive nonnative species policy with Executive Order 13112. This order is binding on all federal agencies and requires them to:

1. Identify any actions they may be taking that promote harmful nonnative species invasions.
2. Prevent the introduction of harmful invasive species.
3. Monitor harmful invasive species (in their lands or areas of responsibility) and take rapid, appropriate action to reduce spread of such species.
4. Restore native species and habitats affected by invasive species.
5. Conduct research on invasive species and develop new technologies for their suppression.
6. Promote public education on the impacts of invasive species.

More broadly, the executive order requires the creation of an Invasive Species Council to serve as an advisory committee at the highest levels of the federal government. The job of the council will be to oversee the implementation and coordination of the executive order within and among the relevant federal agencies. The council will encourage planning at multiple levels of government, provide guidance to agencies on how to prevent or control damage from invasive species, facilitate the development of an information network among federal agencies for monitoring and documentation, establish a coordinated set of databases on nonnative species, and prepare a National Invasive Species Management Plan. An important feature of this plan will be to review existing and prospective approaches and authorities for preventing the introduction and spread of invasive species—that is, to determine if new laws or regulations are needed and, if so, in which areas.

The council and plan offer an unprecedented opportunity to create a coordinated and comprehensive system for preventing and managing nonnative species problems. If the federal agencies responsible for working with the council and implementing the plan follow through on the intent of the executive order, the United States may be able to slow the rate of new introductions significantly.

## Prevention at the International Level

No matter how urgent the need to improve prevention measures, change will be incremental—at the international level even more so than within the United States. Reaching a consensus across the broad range of interests that different countries bring to the table is made all the more difficult by the fact that there does not yet exist a general treaty on species movements (as there are for more long-standing issues such as trade and human rights). A short list of specific, relatively achievable improvements is therefore the most logical place to start.

In a recent article on globalization and invasions, Worldwatch Institute research associate Christopher Bright offered a "three-point agenda" that addressed a few of the most pressing issues related to preventing nonnative species invasions. The first item on the agenda is the development of a legally binding international protocol for ballast water treatment in order to stem the flow of marine organisms between ecosystems. Since the necessary technologies are mostly in place and a regulatory model already exists in the form of the U.S.–Canada prohibition on dumping of ballast water in the Great Lakes, this is a relatively easily achievable goal.

The second item on the agenda is to reform the World Trade Organization's Agreement on the Application of Sanitary and Phytosanitary Standards, known as the SPS. This legally binding measure regulates the measures countries may take to prevent the importation of products that are in any way contaminated (either with chemicals or with undesirable organisms). The SPS is in several respects fundamentally biased towards free trade and against environmental safety. For example, it requires that a country perform a risk assessment before placing any restrictions on trade, even if the risk in question is one of invasion or some other kind of irreparable damage. In Bright's estimation, "the SPS in its current form is likely to make preemptive action [against invasions] vulnerable to trade complaints before the WTO." The SPS also requires that before a country impose any kind of restriction on the importation of a species that is already present in the country, it must demonstrate that the species is a genuine concern by establishing an "official control program" for that species. This forces countries to restrict their prevention efforts to only the most serious threats, essentially guaranteeing that other, lesser problems will only get worse. "Such an approach may be theoretically neat," Bright explained, "but in the practical matter of dealing with exotics, it is a prescription for paralysis."

The third item on Bright's agenda is the construction of a global database on invasions and invasive species. A clearinghouse of this sort would make information on nonnative species issues much more easily accessible to the media and would therefore promote increased public awareness of the threat of invasions. Perhaps even more important, a global system for tracking information related to invasions would be invaluable to scientists and bureaucrats who want to gather information on invasive species before they arrive.

Prevention is a concept that does not come naturally to a society that places full faith in its ability to emerge victorious from battle no matter what the odds. It should come as no surprise that one or another government agency or conservation organization is always declaring a "war on weeds." The American approach to danger and uncertainty has always been to go in guns blazing and sort things out after the smoke clears. Though few would say it in so many words, Americans think prevention is for wimps.

But this is a war that can only get worse. The time has come to admit that there are some battles that are better prevented than fought. With a little luck, the world might learn some ecological humility before the casualties become too great.

## CHAPTER 8

# After All the Sheep Are Gone

## The Recovery of Santa Cruz Island
## after 140 Years of Grazing

The eastern end of Santa Cruz Island loomed a harsh and dusty brown on the horizon as we flew in from the southern California coast. It was July, and there had been no rain for several months. Through 10 miles of hazy midmorning sun, there was little indication that anything was alive on the island.

As we flew in over the eastern shore, the sheep trails that scored the ridges and gullies from the water's edge all the way to the 1,500-foot peaks that dominate the East End came into view. But in the rising late morning heat, the sheep I had come here to see were nowhere to be found. Finally, we spotted a small herd huddled under a clump of low oaks as we banked west along the island's southern coast. Aside from those few scattered trees, though, there was little that still clung to this near-deserted landscape. Viewed from 1,000 feet up, the eastern end of this 96-square-mile island seemed too harsh even for creatures as indestructible as wild sheep.

Most of the year, the whole of Santa Cruz Island is a dry and unforgiving place. It has what is known as a Mediterranean climate, and during the long, hot summers there is little or no rain. But in the winter, the Nature Conservancy–owned western nine-tenths of the island becomes something that even an easterner might call lush, with wildflowers and grasses and gray-green shrubs blanketing the hills and woodland in the draws and on the cool north-facing slopes. But even when I visited in July 1997—at the height of summer—the contrast between the sheep-free Conservancy property and the sheep-infested East End (which had been run as a private sheep-hunting ranch until its purchase by the National Park Service in the mid-1990s) was dramatic.

Our plane crossed the line that marks the beginning of the Conservancy

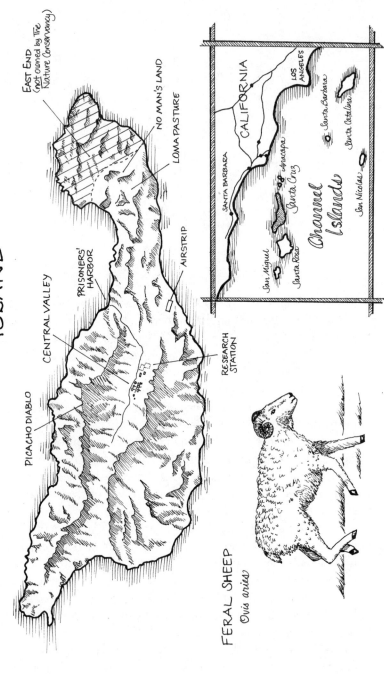

SANTA CRUZ ISLAND

EAST END (not owned by The Nature Conservancy)

NO MAN'S LAND

LOMA PASTURE

AIRSTRIP

PRISONERS' HARBOR

CENTRAL VALLEY

RESEARCH STATION

PICACHO DIABLO

CALIFORNIA

LOS ANGELES

SANTA BARBARA

Santa Barbara

Santa Catalina

Anacapa

Santa Cruz

San Miguel

Santa Rosa

San Nicolas

Channel Islands

FERAL SHEEP
Ovis aries

In 1997, the dividing line between the part of Santa Cruz Island that had been cleared of sheep and the part that had not was clear even from a thousand feet up. (Jason Van Driesche)

property and flew in low over sage-covered bluffs and deep ravines lined with scrub. The air was tinged with a slightly aromatic haze, and the canyon bottoms showed occasional flashes of water. There was life on this side of the fence. It is difficult to believe that when sheep impacts were at their peak in the 1950s, these bluffs had been grazed nearly as bare as the East End is today.

Sheep control was initiated by Santa Cruz Island's previous owners decades before the Conservancy took over management of the island. As a consequence, when the Conservancy began a systematic program of fence repair and sheep eradication in the early 1980s, all but the remotest parts of the island were already free of sheep. Within about eight years, the final Conservancy-led effort had eliminated the remaining animals from the 90 percent of the island it owned, and the process of recovery began in earnest.

A little less than a decade later, I was on my way to the island to take a firsthand look at the changes and to talk with the island's resident ecologists about the challenges that remain. While the removal of sheep has brought tremendous ecological benefits, it has also triggered a cascade of ecological reorganization that has made further restoration anything but straightforward. This is an ecosystem in transition, and no one can be sure exactly where it is going.

We touched down on a dirt airstrip that was surrounded by dense 12-foot-high stands of fennel, a nonnative plant that sprang up across certain sections of the island after the cattle that had also been ranched on the island were removed in the late 1980s. What should have been chaparral and coastal sage scrub was now completely dominated by a garden vegetable gone wild. The release from grazing pressure made no distinction between native and alien plants, and once sheep and cattle were gone, some parts of the island were quickly taken over by fennel and other weeds.

One of the trucks from the research station was at the airstrip to meet us. We threw our bags in the back and piled in. The dirt road back to the station was rutted and rocky, but that didn't stop the man at the wheel from driving fast enough to surprise a wild pig that was trotting along the road. It dove for the bushes and headed upslope as soon as it saw us coming. It seems that the island's nonnative feral pig population is on the rise also, partly in response to the end of the drought, but perhaps because of an increased food supply in the wake of sheep eradication as well.

We pulled into the University of California's Santa Cruz Island Reserve research station about ten minutes later. There were a half dozen or so people already here, most of them working on one or another research project related to restoration of the island's native communities. Their research—indeed, almost all work on the island—is shaped by the 130-year presence and 10-year absence of grazers on the island. The island's ecological history is the most powerful force shaping its future, and any restoration efforts must take into account the fact that sheep and cattle were here.

## The Wasting of an Island Ecosystem

Almost all the ecological changes on Santa Cruz Island—past, present and future—are shaped at least to a degree by the fact that it is an island. Perhaps the most important consequence of its relative isolation is the fact that the island has no large native herbivores, and before about 1850, its plant communities were not subject to grazing and browsing pressure. Chaparral and forest communities were reported to be much more extensive on the island before European settlement than they are today, and vegetation was in general much more dense than at present. In an 1857 account, the U.S. Coast Survey described an area known as Coche Point as being "covered with a thick growth of dwarf oak" and characterized the island's eastern ridge as "a high range of wooded mountains running apparently across the island." A second Coast Survey report filed a year later recounted an attempt to climb a slope above Prisoners' Harbor on the north side of the island that was frustrated by extremely thick, high brush. All of these areas are now mostly grassland, with scattered brush and occasional small trees.

Aside from periodic wildfires, there was nothing in the pre-European ecosystem to keep the island's vegetation from achieving and maintaining maximum growth within the constraints imposed by climate and resources. While the island's Native American inhabitants had for thousands of years played a role in shaping the landscape—they harvested plants, altered some habitats, brought in a few plant species from the mainland, and possibly burned some parts of the island—they did not fundamentally alter the functioning or composition of the ecosystem.

With the arrival of the first Europeans in the early 1800s, several factors came together to force a rapid shift in the composition and structure of the island's vegetation. Several decades of intermittent severe drought and unsustainable harvesting of lumber and firewood on the island contributed to the process of community alteration. However, sheep grazing undoubtedly was (and for many years remained) the primary force driving ecological change. The pressures that sheep exerted on the system forced a fundamental shift of ecological state, altering the context within which all other species live to this day.

Sheep were first introduced to Santa Cruz Island sometime around 1850, and the island's early owners initially found it to be an ideal environment for ranching. There were no predators and no poachers, and the lack of large native herbivores meant that the sheep had virtually no competition for food. By 1857, the island was a major supplier of mutton to the mainland and was considered one of the best-run sheep ranches in the state. Herds were allowed to roam freely and were rounded up once or twice a year for shipment or slaughter. By 1870, the sheep population had surpassed 50,000 and continued to rise despite the fact that tens of thousands of sheep were shipped to the mainland during this period.

This tremendous population explosion in the first decades after introduction was fueled by the accumulated biological capital of an environment that until that point had been free of grazers. The sheep's annual consumption of vegetation far exceeded the productive capacity of the ecosystem, and the crash that came with a series of mid-1870s droughts was inevitable. Fifteen thousand sheep were slaughtered in 1875 and 25,000 in 1877 for lack of feed. But even after such a drop in population, an observer described the island in 1883 as "overstocked" and noted severe erosion in many of the pastures favored by sheep. Three decades of unrestrained exploitation had dramatically reduced the productive capacity of the vegetation, and grazing pressure from the sheep that remained was more than enough to prevent recovery. A downward spiral of degradation had begun.

There is evidence that soil loss on the island accelerated tremendously in the last decades of the nineteenth century and the first decades of the twentieth as continued overgrazing kept the vegetation sparse. Many of the gullies that today scar island hillslopes undercut the roots of still-living trees, indicating that most have formed in the last century or so. In addition, core samples taken in the floodplains of several major canyons and valleys show an abrupt transition in the type and volume of material deposited during floods—and thus of the erosional processes at work on the hills above—that coincides roughly with the initial explosion of the sheep population. Buried below the surface in most canyon bottoms on the island are thin, orderly layers of silt. These are overlain with jumbled rocks and coarse debris, as would be deposited by mudflows, landslides, or other kinds of massive erosion. In general, all but the most recent of these silt layers contain charcoal, which indicates that they were deposited at a time when wildfires occurred periodically. Since historical records indicate that there have been no significant fires on the island since at least the turn of the twentieth century, and since the coarse debris deposits contain no charcoal, it is reasonable to assume that the dividing line between fine and coarse deposits—that is, between moderate and massive erosion—coincides approximately with the arrival of sheep on the island.

The process of mass wasting slowed somewhat across much of the island in the early part of the twentieth century as gullies wore down toward bedrock, forming relatively stable channels that carried water rapidly to the ocean. Soil and vegetation loss was attenuated further still by a change in ownership of the western 90 percent of the island in 1937, when the focus of ranch operations shifted to cattle and the island's owners began active sheep control. Between 1939 and 1980, about 260,000 sheep were either transported to the mainland or shot. This shift in ecosystem dynamics represented a transition to a new ecological state—one in which the relationship between humans and sheep was no longer "protector-protected," but rather "predator-prey." Sheep populations in the more accessible regions of the island were suppressed, and the vegetation in these areas began to recover. However, while this new arrangement significantly reduced the impacts of sheep in some parts of the island, sheep impacts continued unabated in other areas. The overall trend on the island was still one of degradation.

As a consequence, the condition of some of the island's natural communities continued to decline. A 1982 study showed significant loss of Bishop pine in the south-central mountains, attributed entirely to "grazing, trampling, and undermining of roots" by sheep. In places where overgrazing was severe, nonnative annual grasses increasingly outcompeted native perennials. Where sheep were numerous, there was a gradual attrition of palatable native species and replacement by alien and a few native species more resistant to grazing. A study published in the mid-1980s found that "long-term overgrazing by sheep [had] resulted in moderate to severe ecological impacts to about one-half of Santa Cruz Island," with impacts concentrated in the most rugged and remote sections of the island. Sheep showed a strong preference for many endemic island plants, prevented regeneration of many shrub species, and were associated with "a sharp reduction in numbers, species richness, and diversity of birds." The report concluded that "feral sheep have severe negative impacts on the native biota of islands and that endemic plants and birds are particularly vulnerable."

Finally, a demographic study found that the sheep population showed no sign of decline, even in the face of what appeared to be scarce food resources. While the birth rates, survivorship, and overall condition of large mammals normally decline in response to habitat degradation, "feral sheep showed none of these responses. . . . Our data, showing a healthy sheep population thriving in deteriorated habitat, indicated that the continued presence of sheep would lead to further ecological degradation." The collective conclusion of these studies was clear: Significant portions of the island

Sheep impacts on the East End of Santa Cruz Island were so severe by 1997 that entire hillsides were sliding away for lack of any vegetation to hold them in place. (Jason Van Driesche)

were suffering severe damage, and the ecological cost of delay was growing rapidly.

There is a big difference, though, between severe degradation and complete catastrophe. Another study conducted in the early 1980s showed the tremendous recuperative ability of the island's vegetation, given a chance. A researcher who fenced off a small plot in an area that was heavily grazed by sheep found that within a few years, ground cover was much thicker and many native plants absent outside the fenced area had returned. The researcher concluded that "sheep exclosure evidence . . . indicates the possibility of rapid vegetal and soil recovery *if soils are not completely eroded away to bedrock.*" Spontaneous rapid recovery was still possible—but for just how long, no one could be sure.

All evidence pointed to the conclusion that since the introduction of sheep, the island ecosystem had been in an unstable state of continuous degradation. However, ecological changes in the wake of the arrival of grazers—prominent among them the loss of soil and the introduction of other invasive species—had precluded a return to the relatively stable state that existed just prior to settlement. Within the context prescribed by the island's history, there existed a new set of possible states for the future of the ecosystem along a continuum from a relatively well vegetated sheep-free system to a 67,000-acre sheep-chewed bare rock. Where the system fell along this continuum depended more than anything else on the role humans decided to take within it.

The option the Conservancy chose was to take an active role in pushing the ecosystem toward some kind of ecologically desirable state. The first step was clear: Remove the sheep from the system. Little else mattered until the sheep were gone. But in the longer term, the point of restoration is not simply to eliminate what is not wanted. Though the Conservancy clearly saw the removal of the sheep to be one of its primary responsibilities, its overall management objective was stated in positive terms: "To preserve and to protect in perpetuity and to enhance the natural ecosystems, the unique natural flora and fauna, the hydrologic features and the natural aesthetic values of the Island." This statement provided a foundation for a process that has been as much about defining a new ecosystem as about restoring an old one.

The eradication of sheep in particular and the restoration of the island in general soon became part of a grand experiment in adaptive management in which the desired outcome is relatively well defined, but the way to get there is not. What follows is a roadmap of the first few years of the process of transition—detours and potholes and all—and a plan for how to navigate whatever comes next.

## How to Kill 31,871 Sheep

The elimination of sheep from Santa Cruz Island did not begin with the first sheep shot by a Conservancy hunter. The Stanton family, who had owned most of the island for half a century prior to the arrival of the Conservancy, went to great lengths to clear sheep from the more accessible portions of the island, building fences and selling or shooting hundreds of thousands of sheep over about five decades. However, even with the head start that the Stantons' work afforded, the Conservancy spent several years preparing for the eradication program in an effort to minimize the time and effort required to clear the island of sheep.

The first stage was a series of research projects that were initiated in 1979 with the goal of answering two questions: What were the basic ecology and present status of the feral sheep population on Santa Cruz Island? and What impacts were sheep having upon the natural resources of the island? The purpose of the research was not only to provide a baseline of information for the control program and for subsequent management; documentation of the impacts of the sheep population would also be necessary should the Conservancy be forced to defend its management actions in court. Even though eradication was urgently needed, the time spent on research was worth the investment.

Concurrent with the research program, the Conservancy began to repair the network of fences across the island that the Stantons had built and maintained as part of their ongoing sheep control effort. These fences divided the island's nearly 100 square miles of terrain into twenty-three manageable-sized pastures that could be cleared of sheep one at a time. The Conservancy investigated a number of options for sheep removal, including trapping and transportation to the mainland. Given the rugged terrain and remote location of the island, though, it would have been prohibitively expensive and difficult to remove live sheep from the island. Shooting the sheep was the only way to eliminate them from the preserve before they made the rest of the island's soil wash into the sea.

Conservancy-sponsored hunting began in the smaller pastures in late 1981 with teams of three to four hunters. The goal of these early hunts was not only to clear the sheep from the smallest, lowest-density pastures, but also to develop techniques that would be effective in the larger pastures with many more sheep. Teams eventually grew to about eight to ten hunters, and each member was equipped with a high-powered rifle, a handheld radio, and enough supplies for a full day in the field. A team would locate a flock, surround it, and pick off sheep one by one as they tried to escape. The hunters made sure to leave no wounded sheep in a given pasture before moving on to the next and rotated among a group of pastures over a series of weeks until all the sheep were dead. For the most part, carcasses were left where they fell, and scavengers-mostly feral pigs—quickly took care of the remains.

By June 1989, there were no more than five sheep known to remain on the Conservancy-owned western 90 percent of the island. The Conservancy's hunters had killed a total of 31,871 sheep over the seven and a half years of the eradication program. The last few sheep were tracked down and shot over the next couple of years. So long as the boundary fence between the preserve and the still-infested East End held, the single most powerful force shaping the island for over a century was gone. This was a new ecosystem.

## The Consequences of Eradication

When I visited Santa Cruz Island in 1997, it had been nearly ten years since the last section of the preserve was cleared of sheep. In many ways, the recovery of the island's native vegetation was remarkable and represented a significant contribution to the conservation of California's native diversity. The natural communities found on Santa Cruz Island are among the most threatened in the nation, for their mainland California equivalents are being converted to subdivisions and shopping centers at an overwhelming pace. What's more, the island harbors many unique species found nowhere else on earth. But while no one would deny the large net ecological benefit of sheep eradication and of island restoration efforts in general, there have been a few undesirable consequences of the last ten years of management.

The view from the plane on the flight in from the mainland, however, would have made anyone give alien species eradication a heartfelt endorsement. If getting rid of sheep meant the difference between bare soil and lush coastal sage scrub, there really seemed to be nothing but good that could come of the eradication of grazers. So as soon as I landed, I arranged a trip to the East End to see the recovery up close.

Rob Klinger, the preserve ecologist, lent me an intern named Brian Kitzerow for the afternoon, and we hopped in a truck and headed out. Brian spent most of his time out along the border with the East End. His job was to patrol the area for sheep that managed to get through the fence and chase them back into No Man's Land, the Conservancy-owned buffer zone between the area that has been cleared of sheep and the still-infested East End. (Once the eradication campaign was completed in 1989, the Conservancy decided for public relations and safety reasons against ongoing use of shooting on the

Brian Kitzerow, a Nature Conservancy intern on Santa Cruz Island, patrolled the border of No Man's Land to keep sheep from reinvading the 90 percent of the island that in 1997 was sheep-free. (Jason Van Driesche)

island.) We left the truck on the ridgetop that marked the border, crossed the fence into sheep territory, and walked down the line until we found a sheep trail that cut down the slope into the boulder-strewn ravine below.

It soon became apparent that Brian was himself at least part wild sheep. I followed him down a network of impossibly narrow sheep trails skirting back and forth along the fault lines of a slope that had more in common with a cliff than anything I'd ever hiked. We picked our way down through the dust and loose rocks, then worked back up the other side to a ridge where Brian had spotted a few sheep as we drove in. The slope was cut with ravines 10 or 15 feet wide, forcing us to choose a ridge between two gullies and head straight up. We passed a few small clusters of oaks about mid-slope. The roots of many of these trees had become stranded in midair as the gullies deepened on either side. There was almost no grass until we reached the upper portion of the ridge, but even then there was more dirt than vegetation.

The sheep had been here recently, though, for there were fresh hoofprints

An ever-wary flock of feral sheep scrambles up a steep slope on Santa Cruz Island. (Peter Schuyler)

in the dust. We followed the tracks along the ridge and stumbled upon the flock on the other side of a small rise. About a dozen sheep were busily trimming the leaves from a group of small oaks. They saw us coming and ran for the next ridge. We stood there for a moment on the naked slope, then turned and made our way back to the truck.

Brian and I continued down along the Loma Loop, a rutted four-wheel-drive track that circles through the coastal sage scrub of the Loma Pelona pasture. This tract adjoins No Man's Land across the southern half of the border and is, according to Rob Klinger, one of the two or three most spectacular recovery sites on the island. Native bunchgrasses were abundant, and the scrub grew 5 feet tall. The road ahead wound down through a small valley and then up the ridge on the far side. We stopped for a moment to take a few pictures. The sagebrush shone silvery green against the gold of summer grasses. There were still gullies—the effects of a century and a half of overgrazing don't disappear overnight—but their edges were softened and their contours made more gentle by the vegetation. The air was distinctly cooler here than on the barren ridges of No Man's Land. Tendrils of fog crept up from the ravine below, wrapping the trees in a faint blanket of white.

Suddenly Brian shouted, pointing to the opposite slope. A lone sheep was browsing hungrily, pausing every few seconds to bleat for the flock it had left

The recovery of native vegetation following sheep removal has been spectacular in the Loma Pasture—an area that a few decades before was grazed nearly as bare as the East End was in the 1990s. (Jason Van Driesche)

behind. It seemed torn between eating as much as it could and getting out of sight. As we drove closer, caution prevailed and the sheep dove for cover. Brian made a note of the location. He would be back here tomorrow to herd the sheep back across the line.

As we drove from Loma back down to the Central Valley, the native vegetation began to give way to patches of 12-foot-high fennel. By the time we reached the valley floor, the road was surrounded by nothing but this introduced invasive weed. The valley was a sea of yellow to its margins, and everywhere the scent of fennel was heavy on the air. Recovery here has meant something quite different than in the surrounding hills. While native plants have made a decisive comeback in the Loma Pelona pasture, nonnative weeds have taken over in the Central Valley bottomland.

Fennel was accidentally introduced in the 1850s, probably along with sheep or cattle brought to Santa Cruz Island by early European settlers. It turned out to be particularly well adapted to the island's Mediterranean climate and to the summer fogs and established easily in and around the Central Valley and the other areas where settlers later planted it as a crop. By the time The Nature Conservancy took over management of the western 90 percent of the island, fennel was widespread but didn't appear to be very abundant across most of the cattle ranch in the Central Valley. It was low on the list of priorities, for while it was clearly well established, it did not appear to be particularly invasive or dominant. It was simply one of the many nonnative plants that formed a backdrop to the more pressing problems of sheep, cattle, and pigs.

As the sheep eradication campaign was winding down, the Conservancy turned its attention in early 1988 toward the cattle. While they were restricted to the less rugged pastures and therefore were a less urgent problem than sheep, the cattle were identified as a fundamental barrier to recovery of native species in the areas where they grazed. In the spring and summer of 1988 the cattle were rounded up and shipped to the mainland, with the expectation that once the Central Valley was released from grazing pressure, native species would make as strong a recovery there as they had already begun to do in the higher pastures.

The island was in the middle of a drought when the cattle were removed, so nothing much changed until heavy rains came in March 1991. What happened then took everyone by surprise. Unlike in the Loma Pelona pasture· and other more remote parts of the island where native species had already begun to appear and proliferate over the last decade, the Central Valley was suddenly filled with fennel. And by the middle of that year, Rob Klinger

Following the removal of sheep and cattle from the island, nonnative fennel plants have sprung up in dense stands over about 10 percent of the island. (Jason Van Driesche)

explained, "It was clear that something drastic was happening—in an ecological sense, something incredibly profound."

In retrospect, Rob said, it was clear that the importance of the different land-use histories of the Central Valley and the surrounding hills had not been fully taken into account. It turns out that the worst patches of fennel were located in areas with deeper soils and more soil moisture, areas that had been plowed and planted in the past. In the years after crops were largely abandoned and the island was dedicated to cattle ranching, these also tended to be the places where "grazing animals created the most disturbance: holding areas, watering holes, and the flatter areas throughout the Central Valley." This means that cattle had both promoted and suppressed fennel wherever they went, spreading the seed and creating an ideal seedbed but at the same time keeping fennel abundance low through constant grazing pressure. What no one had realized was just how fundamentally this dynamic would change when grazing pressure was released.

While the Conservancy had considered in its eradication plans the possibility that nonnative plants would increase in abundance once grazers were removed, this possibility was unfortunately not given enough weight. The preserve's managers were largely operating on the unspoken assumption that—as Rob Klinger and two other ecologists noted in a study of the changes in vegetation after sheep removal—"removing nonnative grazers will lead to recovery of native species." What in fact happened was that removal of grazers led simply to the recovery of whatever vegetation was present on a given site. Implicit in the eradication campaign was the assumption that history did not matter; that once sheep and cattle were removed, the island would be a blank slate, and nature would write upon it as the "original" native script intended. The reality, of course, was that history is essential, and there is no intention in the composition of natural communities. While removal of grazers did indeed spur the recovery of the island's vegetation, the composition of that vegetation had been fundamentally altered long before anyone even considered removing feral animals from the ecosystem.

On the way back to the research station, Brian took a detour on a road that led to the north coast. I wanted to explore the pine forest and chaparral that blanketed the cooler, moister north-facing slopes of the island, and Rob had suggested the hiking trail that ran from Prisoners' Harbor at the mouth of the wash to Pelican Bay a few miles down the coast. Brian let me out at the water's edge, and I started up the trail. Here, too, the vegetation had made an impressive recovery. Tufts of bunchgrass covered the ground in a dense blanket, and the shrubs were nearly as tall as I was. There were quite

The recovery of native vegetation has been spectacular on many parts of the island.
(Jason Van Driesche)

a few young trees on the higher slopes, most of them less than ten years old.
The land had a feeling of exuberance to it—of a place that had just been
relieved of a tremendous weight.

But there was fennel along the north coast as well—only a few plants
scattered across the hillside, but their presence was worrisome even so. Why
was it here where there had been almost no cattle? Did the dynamics of the
current system favor its continued spread, or would it stay put until some
effective control measure could be found? This north slope had been particularly heavily grazed by sheep, and now it clearly was an ecosystem in transition. The problem was, no one really knew where it was going.

Rob Klinger compared the island ecosystem before the removal of grazers
to a set of interlocking coiled springs. "When you release one," he explained,
"they all take off in different directions." Each place on the island reacted
differently to the eradication of sheep and cattle, some in ways no one had
expected. In response to the fennel crisis, the Conservancy refined its overall approach to invasive species management on the island. Consequently,

while removing grazers was undoubtedly a net positive change for the island, the next round of management had the potential to be better still.

Recognizing the fact that "the removal of [a nonnative] species will have at least as profound an influence on the system as the original invasion of the species did," a study of the expansion and control of fennel on the island suggested that any nonnative plant management program should begin by studying three fundamental processes:

1. The reasons underlying the successful establishment of the nonnative species
2. The effects of the nonnative species on the communities in which it occurs
3. The probable effects of removing the nonnative species on the communities it has invaded

This approach provides a mechanism for shifting attention from a negative single-species target to a positive ecological goal. The Conservancy was still keeping an eye on history, but it had begun to focus in earnest on the island's future as well.

The conceptual foundation underlying this fennel research was a state-transition model in which as many states as possible of the system under study are defined and the processes that are thought to result in a transition from one state to another are spelled out. This kind of model explicitly rejects the traditional ecological paradigm of succession toward a single climax in which all pathways of change are assumed to converge on a common vegetation type. In so doing, it acknowledges that removing or controlling an invader will not necessarily induce a return to a preinvasion state, for the invader may have pushed the system into a transition toward an entirely different state. In this case, the island ecosystem may have already been too far along in a transition that led to domination by fennel to allow for spontaneous resurgence of native vegetation, and moving the system toward a native species-dominated state probably required more vigorous intervention than simply removing grazers.

As of 1999, the management strategy for the long-term control of fennel called for a combination of herbicides and prescribed burning, an approach that appeared to be working. However, while the fennel explosion certainly constituted a major ecological crisis, the most pressing alien species problem remaining on the island was not fennel, but feral pigs. While fennel was found only on about 10 percent of the island, pigs roamed freely. Pigs are aggressive generalists that are able to make use of a wide range of resources and have an impact on a wide variety of community types. They root the soil and trample vegetation in search of roots and tubers. They also consume large quantities of acorns and fruits, all of which are important food sources for some native island animals (especially birds). As such, they competed with native island animals for food, prevented regeneration of many native plants, and promoted the spread of invasive nonnative plant species, forcing progressive alteration of natural systems. Most important, while fennel's abundance appeared to be largely the product of past damage caused by grazers, pigs were themselves agents of nonnative disturbance and as such altered the system in ways that are not favorable for native species. The fight against fennel (and against most other invasive plants) would be far more successful once pigs were gone.

## The Next Eradication

Pigs have been present on Santa Cruz Island since at least the mid-1850s. One of the earliest accounts of their introduction tells of a sea captain who turned a few pigs loose on the island around 1854 in order to ensure a supply of pork for passing sailors. It is likely that the pig population grew throughout the end of the nineteenth century as it expanded its range across the island. By the beginning of the twentieth century, the pigs had become "so numerous as to be an intolerable nuisance," and the owners of the island offered a bounty of five cents to their workers for each pig they killed. Given the island's rugged terrain and the pigs' tremendous reproductive ability, this early extermination campaign probably had little effect on the population. However, as the island became more and more denuded from continued overgrazing by sheep, the pig population may have begun to decline as degradation of the ecosystem as a whole reduced the availability of food.

By the time The Nature Conservancy was wrapping up the sheep extermination campaign, the pig population was estimated at somewhere between 2,000 and 4,300, which averages 3.5 to 7.5 pigs per square mile. The population was not spread evenly over the island, though, for the pigs tended to avoid high, rocky ridges and areas where humans were present. As vegetation recovered and food supplies increased in the wake of sheep eradication, the pig population continued to rise. By about 1994 the population peaked—as Rob Klinger put it, "there were just a lot of pigs out there." But in 1995, a poor acorn crop combined with classic overshoot of carrying capacity triggered a population crash, and their numbers fell to fewer than a thousand.

Unfortunately, the Conservancy was not in a financial or a logistical position at that point to take advantage of the drop in population and mount an eradication campaign. However, eliminating pigs from the island remains the Conservancy's highest priority, for as Rob explained, "pigs are the single most significant environmental issue that we are confronted with out here. Nothing else we do will matter until we get rid of the pigs." The management plan in preparation for the island calls for eradication of the growing pig population using shooting, snares, traps, or some combination of the three. Erik Aschehoug, an ecologist working on the plan, is confident that pigs will be gone by about 2004.

Eradication can't come too soon. It turns out that pigs were the cause of the patches of rooted-up ground I'd seen along the trail above Prisoners' Harbor. The pig population along the north shore was growing, and the signs of damage were everywhere. Rob was afraid that pigs might provide the disturbance that fennel needs to keep expanding its range beyond the Central Valley and the few other areas where cattle had been pastured. There is as yet no direct evidence for a link between pig presence and fennel invasion, but the ecology of the two species appear to complement each other.

The next day, I decided to hike up Picacho Diablo, the highest peak on Santa Cruz Island. There was a trail to the peak that began at the west end of the Central Valley. The first few miles switchbacked across the face of a long, grassy slope, and there was nothing to block the view for at least a mile or so across the ridge. Before long I saw a group of pigs—far off in the distance, but clearly recognizable. They were moving fast, stopping only every few hundred yards to sniff the air and scan the horizon. There wasn't much for them to eat here. Chances were good they were heading for the other side of the ridge, where vegetative cover was thicker and food more abundant.

When I got to the top and looked down the north slope, pig sign appeared to be more common. The pigs were active in one of the areas that had recovered most fully to native vegetation after the removal of sheep. If pigs remain there very long, the disturbance they create might tip the balance toward nonnative plants even in areas now dominated by native species. But just as sheep eradication and cattle removal caused some unexpected changes, elimination of the feral pig population is sure to bring some surprises. With the lessons of sheep and cattle removal in mind, the Conservancy and the National Park Service (which now owns the other 10 percent of the island) are designing a pig eradication strategy that will attempt to identify and dampen the reverberations that removal of pigs might cause. The development of such a strategy begins with the three questions posed by the study of fennel expansion and control mentioned earlier: Why were pigs able to establish successfully on the island? What impacts do they have on the communities in which they occur? And what are the likely effects of pig removal on the natural communities in which they are found?

Substantial progress has already been made on the first two questions. Addressing the third may take the form of exclusion experiments, analysis of similar efforts in other places, or simply a systematic consideration of possible changes in ecological dynamics in the island's various natural community types. It is unlikely that pig removal will have as drastic an impact on nonnative species populations as did grazer removal, since pigs—in contrast to sheep and cattle—don't seem to be suppressing any major nonnative plants. Even so, an effective strategy would need to have as its theoretical foundation a state-transition model similar to the one developed for fennel control in which the ecological relationships and transitions between the various possible states of communities now driven by pigs are mapped out as carefully as possible beforehand.

The experience of sheep eradication and cattle removal has made it clear that alien species control is a tool for native ecosystem restoration, not an end in itself. While control efforts are essential, even more important is to develop both a specific restoration goal and a map of overall ecosystem dynamics. What ecosystem state or complex of states is most likely to sustain the full scope of native diversity over the long run? How are these desirable states related ecologically to less desirable states, and what triggers a transition between them? Unless alien species control is conceived and carried out within the context of restoration of native ecosystem function, the effects of control will persist only so long as we keep our hands on the wheel. In the long run, the system will have to function without human intervention, and the more we understand of its dynamics, the better equipped we will be to point it in a direction that most benefits native species—and let it go.

## Thinking about Adaptive Management

With this long-term perspective in mind, the Conservancy and the National Park Service (which now owns and has begun to manage the East End) began working together in the late 1990s to create an adaptive management strategy to guide restoration on Santa Cruz Island into the next century. Rob Klinger, the Conservancy's preserve ecologist, invited me over to the research station one evening to discuss the island's future. His approach to management in a constantly changing environment distills down to three basic principles: perseverance, realism, and flexibility.

"We operate in a society that watches half-hour television shows, expects a sporting contest to be wrapped up in one to two and a half hours—everything is compressed," Rob explained. "We are a society that has a compressed perception of time. We carry that over into ecosystem management and conservation. The changes that we are going to see are extraordinarily complex, and they're going to be functioning at different rates according to the species or community that you're working with. And that is a huge, huge challenge."

The Conservancy has been doing restoration work on the island for over twenty years, and the ecosystem is only just beginning to settle into a reasonable range of variability. A few days before, I'd been looking at some early photos of the island with Dr. Lyndal Laughrin, the manager of the University of California–Santa Barbara field station who has been conducting research on the island for three decades. Dr. Laughrin has a longer history on the island than anyone alive, and he understands better than anyone the magnitude of the changes this ecosystem has undergone. We had walked outside the field station library and looked at a hillside shown in a photo from the early part of the twentieth century. A full ten years after the removal of the grazers that had destroyed the presettlement chaparral, the slope was still covered in little but grass. Damage of this magnitude took over a century, and restabilization may take nearly as long.

"Even in the most natural of areas," Rob emphasized, "when you have residual effects, you're going to have to be prepared to do management over the long term." Fixing an ecosystem is not at all like fixing a clock, he explained, where a spring or battery or a quick cleaning makes it as good as new. Ecosystems can't really be repaired so much as coaxed. "There are no quick fixes," Rob continued. "We have to remember that we're only ten years now beyond the end of eradication, after a hundred and fifty years of grazing. It's important to recognize that we're making incremental changes. It's taken such a long time to get here, and our knowledge of how these systems operate is so imperfect that we're going to have to learn as we go."

It was getting dark, and we sat looking out at the sharp line of the evening sky. All around the research station were enormous old eucalyptus trees planted by the first European settlers on the island. Rob had told me that these were some of the oldest eucalyptus trees in California. Even though they were nonnative, their historical significance precluded their removal from the island. Rob reluctantly accepted their permanence and had learned to work around them.

"We're not going to be able to get rid of all the exotics," he said emphatically. "I can only think of a few species that we will be able to eradicate completely. That's still going to leave twenty-three percent of the flora as nonnative. The long-term goal is that, yes, they will be a part of the flora out here, but the key is that they will not be a significant component, and they will not be driving the system.

"The idea here is that in probably any natural area—not just this island—you have to recognize the fact that you are going to continue to get invasions, and that you are going to have to manage residual effects from what is already here," he continued. "If you can maintain or increase the native species diversity, and maintain a functional ecosystem that leads to that level of diversity, you've accomplished something enormously significant."

The evening air was cool. After a while, we moved inside and sat down around the main table in the research station bunkhouse. On the wall was a map of the island with large areas marked off with hatch marks or dots or slashes. This was the command center for the fennel control program, and each kind of mark on the map represented a different kind of management experiment. Rob wanted to know what was going on, and this was one way of keeping track.

This kind of monitoring, Rob said, is an absolutely essential part of any program of nonnative species management. "Aside from the actual removal," he added, "it is the most important thing you can do." Without a monitoring program, it is altogether too easy to get caught up in invasive species control for its own sake and forget that the real goal is recovery of native communities. "Monitoring is really the only way you're going to be able to evaluate the effectiveness of what you did," he said, "and what you might need to do in the future.

"These systems are always changing, and our perception of what should be here is probably simplistic at best and inaccurate a lot of the time," Rob said. "We have to try to avoid that bias. In our efforts to manage these lands, we can actually overmanage, and we can do inappropriate things, simply because the land is going to take its own trajectory. We don't know enough about systems interactions between the species, how communities relate to each other, or the cycling of nutrients to even begin to think we can start manipulating these things. So for our own sanity—and for efficient use of resources—I think there's going to have to be a significant monitoring component to all this, where we just step back and study some areas and learn."

It was late. I headed back along the road to the field station through rows of eucalyptus and tangles of fennel and rustlings that might have been pigs. This place has seen some hard times, I thought. But then, so have most places. The difference here is that while the degraded state of parts of the island is acknowledged, it is not accepted as immutable. This kind of realistic optimism is what has made restoration work on the island so successful.

It is why, at the close of the twentieth century, plants that haven't been seen in decades are reappearing on the island and native trees that for decades only lost ground are now growing into forests once again.

Underlying this approach to restoration are the three basic principles—perseverance, realism, and flexibility—that kept coming up whenever Rob talked about long-range management. On the way back to my bunk, I thought about how these ideas might apply to handling invasive species problems generally. What they boiled down to was this:

- *Be prepared for the long haul.* Fighting alien invasive species is a process of experimentation that is most effective when it is geared not only toward short-term victories but also toward long-term recovery.
- *Learn to see things for what they are.* The most effective way to influence the dynamics and composition of any ecosystem is to acknowledge and work with the fact that, in most cases, nonnative species are a permanent presence in the system.
- *Work with the fact that everything changes.* Staying abreast of the workings of the system and the roles of its various member species (before, during, and after any management action) is the only way of keeping strategies pointed toward goals.

Unfortunately, this approach is not cheap. Nor is it easy. But in the long run, this is the only approach that makes any sense, economically or ecologically. Within the context of overall ecosystem restoration, a carefully designed program of invasive species control should eventually reduce the level of human intervention that is necessary to perpetuate and strengthen native communities to the point where, as Aldo Leopold put it, we humans are just a plain member and citizen of the land.

The air was clear and cloudless the afternoon I left the island. We roared off the airstrip in a swirl of dust, cleared the bluffs, and climbed out over the ocean. Even from a couple thousand feet up, the island was too massive to take in all at once. The southern shore curved out of sight past Willows Anchorage. The floor of the Central Valley was visible only in patches between the peaks of the East End. Deep shadows filled the ravines that drained out of No Man's Land. There was a lot of room on this island for surprises to hide, and rewriting the ecological score on this grand a scale is the height of ambition. At this point, though, there really is no other choice.

Since I was on the island in 1997, the Conservancy and the National Park Service have taken some enormously positive steps toward long-term stabilization of the island. Most important, the Park Service eradicated the remaining sheep from the East End (including a few that had gotten through holes in the fence and reinvaded the preserve). Once pigs are eliminated from the island as well, its future will be a good deal more secure.

There are no doubt more surprises to come. But the most important step in designing the ongoing recovery of the island is to acknowledge that all ecosystems are always less than entirely predictable. And the next step is to do everything possible to see the surprises coming and—if the surprises turn out to be unpleasant ones—dampen their effects before they take the system off in a direction that will only amplify the effects of past damage. Just as important, of course, is to stay alert for the good surprises and encourage them as soon as they become apparent. The path of recovery on Santa Cruz Island will be shaped largely by the ability of the people charged with taking care of it to adapt to whatever opportunities and challenges the island presents.

This is the shape of conservation from here on out—juggling past mistakes with present knowledge in the hope of coaxing something better out of the future. Like it or not, we have imposed upon ourselves for many generations to come the role of stewards of the earth. Santa Cruz Island is as good a place as any to begin to learn how to carry that responsibility well.

CHAPTER 9

# Holding the Line

## *Chemical and Mechanical Control*
## *of Nonnative Species Invasions*

Shooting thirty thousand sheep is a messy business. Spraying hundreds of acres of fennel with herbicides ought not be done lightly. But in an age of invasions, protecting nature sometimes means destroying nature that does not belong. Machetes and rifles and herbicides and poison baits and all the other forms of mechanical and chemical control can be extremely effective tools for nature conservation—but only when they are used with care in situations where such methods are appropriate.

Unlike biological control methods, these approaches can be restricted to particular places, making it possible to control a species in one area and leave it unaffected in another. For example, a park may be fenced and goats exterminated within the park while goat herding continues on adjacent private lands. Counterbalancing these advantages are several obvious drawbacks to chemical and mechanical controls. First, herbicides, traps, fences, and labor are expensive and become prohibitively so as the size of infested area to be treated increases. Second, killing invasive animals (especially large and furry ones) can stir up a lot of controversy, even if failing to control them does far more harm to native species of animals and plants. Third, herbicides can be dangerous and environmentally damaging, especially if used improperly. And finally, some invasive species cannot be controlled at all with chemical or mechanical techniques. With these inherent limitations in mind, this chapter examines the appropriate uses of chemical and mechanical controls as tools for protecting natural areas from invasive nonnative species.

## Chemical and Mechanical Control of New Invaders

Many nonnative species that ultimately become damaging in natural areas were introduced deliberately. They are not so much invaders as invited guests that have become obnoxious and refuse to leave. Lehmann lovegrass, for example, was introduced to Arizona in the 1930s after a worldwide search for grasses suitable for revegetating overgrazed, degraded native grasslands in southern Arizona and was widely promoted by the U.S. Soil Conservation Service for years. It has since proven highly invasive and damaging to grassland communities and has replaced native grasses in Arizona over several hundred thousand acres. For species like this one, the concept of early detection and eradication is irrelevant, and chemical and mechanical controls are of limited use because such species are seen as pests only long after they have become irrevocably established and widely distributed.

In other cases, though, chemical or mechanical control might be useful for slowing the continued spread of newly introduced invasive species if the public realizes that the species is damaging before it becomes widespread. In such cases, herbicides and chainsaws can serve as a delaying tactic after the invasion has occurred, but before all vulnerable areas have been reached. For example, the invasive tree *Miconia calvescens* was introduced to Hawai'i in the early 1960s and was extensively planted for its ornamental beauty. However, once scientists recognized its ability to seed within and overwhelm native forests, the invitation was rescinded. A statewide campaign has since been organized to locate infestations on other islands, and a major control and eradication effort is under way. With early detection, the smaller miconia infestations on some islands might still be eliminated by cutting and application of herbicides (see Chapter 11).

A rapid response system like the one set up on the Hawaiian island of Maui to handle new invasive species threats—first the miconia crisis, then a wider variety of alien species problems—would be the most effective means of dealing with new or spreading invasions in any context. Such a system would have three main components: critical watch lists to focus attention on established species that are likely to become invasive, monitoring to detect any sudden expansion of established species before they become widespread, and strike teams to control or eradicate such species quickly and efficiently. If this kind of approach were adopted nationwide, response would have to be coordinated at least to a degree by a national interagency body in order to ensure complete coverage, but implementation would probably fall to specific land management agencies or affected states or regions.

## Critical Watch Lists

If government agents or citizen watchdog groups are to protect natural areas effectively, they need to know which nonnative species pose the greatest threat to the integrity of native ecosystems. Because hundreds of new non-native species still become established in the United States every year, it is not possible (nor is it entirely necessary) to control all new arrivals. While there is no sure way to determine if a given species will become a damaging invader (see Chapter 5), certain species are more likely than others to prove invasive. A research project begun in 1999 by USGS scientists at Haleakala National Park on the Hawaiian island of Maui is attempting to identify potentially troublesome species by comparing inventories of nonnative species present but not yet invasive on Maui with lists of species known to be invasive elsewhere in the world. Those species that appear on both lists would be targeted for careful monitoring and possible control. While this approach could never identify all possible invaders, such a list would help focus control efforts where they are most needed (see Chapter 13).

## Monitoring Established Species for Invasiveness

For species already well established but not yet obviously damaging, regular monitoring of abundance in natural areas is essential for detection of sudden increases in density. Many introduced species have a latent period during which their population level is so low that they go unnoticed. The purpose of monitoring is to detect when such species become abundant enough to be considered invasive. The above-mentioned research project on the island of Maui includes a monitoring component as well and will collect detailed base-line data on the distribution and abundance of alien species of concern and then monitor those species every year for significant range expansion or other indicators of invasiveness. This kind of information will be useful for identifying and controlling species that are just starting to increase in abun-dance before they become so widespread as to require expensive long-term control.

## Strike Teams for Rapid Control

Once a newly invasive species is detected, mounting a rapid response may make the difference between a relatively inexpensive program of eradication and a drawn-out battle of suppression. Many species that eventually become invasive show an exponential population growth curve—much like that of the earth's human population—in which a constant rate of growth at first produces an insignificant increase in absolute numbers, but eventually causes what seems like a sudden population explosion. Response will be effective only if it comes at the point where an invader's numbers have just begun to increase rapidly, instead of waiting until the population has become unmanageably large. New invasions are like house fires, and just as a fire department's only responsibility is to prepare for and respond to fires in order to prevent them from spreading, a strike team must consist of trained, well-equipped people who have no responsibilities other than to respond immediately and vigorously to invasive species threats.

Success or failure hinges not only on the existence of an on-the-ground strike team, but also on clear and coordinated policies for eradication or control. Only a state or country that has a comprehensive policy on nonnative invaders and what to do about them will be able to act in time to eradicate or easily suppress a new invader. New Zealand now has well-developed protocols for responding to new invasions, and Maui and other Hawaiian islands are in the process of developing similar systems. However, most countries and states do not have any such measures in place. In most of the United States, invasive species spur a reaction only if they are pests likely to threaten agriculture, public health, forestry, or recreation. Invasions of new plants or other species that are potentially damaging to native ecosystems but not to economic resources generally go without response until the invader is widespread and quick suppression or eradication with chemical or mechanical means is impossible.

## Chemical and Mechanical Control of Widespread Invaders

Even if an invasive pest species becomes widespread, chemical or mechanical controls may still be useful in some contexts. Whether these are appropriate methods in a given situation depends on two factors. First is the feasibility of control. Some species simply are too small, or too hard to catch, or too numerous for poison baits or machetes to be worth using. Second is the potential of control to catalyze positive long-term change. Even if chemical or mechanical controls significantly reduce a pest's population level, there generally isn't much point to spending money on herbicides or shooting year after year if a few missed seasons would allow the pest to return to its precontrol level. However, there are some circumstances under which long-term chemical or mechanical control is justified, particularly when the invaded ecosystem is irreplaceable and when continuous control is an effective means of reducing invader impacts. A well-considered decision to use some chemical or mechanical method of control has to take all of these factors into account.

### *Feasibility of Control*

A number of factors affect the feasibility of control. Some have to do with the biology of the invader itself; others are intrinsic to the invaded environment. Both elements are generally at play, though, in determining how good a target any particular species might be for chemical or mechanical control.

One key issue is the *form and size of individuals of the species to be controlled* in relation to the nontarget species in the invaded environment. If invaders present themselves as large, discrete individuals rather than as small individuals intermingled with other species of similar size and form, they may be controllable or even eradicable over large areas. For instance, crews with chainsaws, machetes, and squirt bottles full of herbicide have been able to clear hundreds of thousands of acres of Australian paperbark trees from the sawgrass marshes of the Everglades because the tree grows mostly in discrete stands and is quite easy to spot above the marsh. On the other hand, if a nonnative grass were to invade the Everglades, chemical and mechanical control might quickly become impossible. Similarly, feral sheep are relatively easy to control or even eradicate, but invasive ants are almost impossible to control on anything but a very local scale, particularly if there are native ant species in the invaded environment.

Another factor is *how easy an individual of an invader species is to kill*. Plant species that regenerate from fragments, such as bits of roots, stems, or parts of floating plants, are especially difficult to suppress with chemical or mechanical methods. In general, control is worthwhile only if the invader

Crews from the South Florida Water Management District killed these invasive paper-bark trees with machetes and herbicides, the first step in the recovery of native Everglades vegetation on the site. (Jason Van Driesche)

In mature stands of paperbark trees in the Everglades, nonnative species control resembles a logging operation. (Jason Van Driesche)

This dike marks the dividing line between an area that has been cleared of paperbark trees and one that has not. (Jason Van Driesche)

François Laroche, a staff environmental scientist with the South Florida Water Management District, explains to Tracy Feldman how chemical and mechanical control methods were used to clear paperbark trees from this sawgrass marsh. (Jason Van Driesche)

has completely taken over an area, in which case extreme measures (such as bulldozing or repeated broad-spectrum herbicide applications) might be justified given that no native species are left to be harmed by control. For example, the Bosque del Apache National Wildlife Refuge uses bulldozers and giant plows to remove both stems and roots of the saltcedar bushes that have formed pure stands across much of the refuge. (Saltcedar is a large Asian shrub that resprouts vigorously unless the entire root mass is unearthed and burned.) Once the ground is entirely cleared of saltcedar, refuge staff flood the area to drown out new saltcedar seedlings and eventually replant with native willow and cottonwood.

A third element to consider is *whether it will take only one or many repeated treatments to eliminate* a stand of a given invader species. Some plants reseed rapidly, either from large seed banks present in the soil or from rapid movement of seeds from distant sources. Most such plants have either light, highly dispersive airborne seeds or seeds in fruits that attract animal dispersers such as pigs or birds. If such plants can be controlled at all, they require multiple treatments over successive seasons. For example, one reason that invasions of the miconia tree in Hawai'i are so difficult to control is that a mature miconia tree produces up to 10 million seeds every year in fleshy fruits that are eaten and spread by animals. While removing a stand of a few hundred mature miconia trees is relatively straightforward, the millions of seeds in the soil make complete eradication from the area a difficult and drawn-out process.

Finally, the *ecological sensitivity or conservation value* of the invaded habitat can be a significant factor in determining the usefulness of chemical and mechanical controls. For instance, invasive aquatic plants generally should not be controlled with herbicides because aquatic ecosystems and species are often highly sensitive to chemical treatments. Similarly, many commonly used chemical and mechanical methods may not be appropriate for controlling a nonnative species that has invaded a rare habitat type or an area with endangered species. As a consequence, some of the most difficult species to control are those that both occur in a sensitive habitat and are themselves difficult to eradicate. For example, water hyacinth, which regenerates easily from fragments, is an invader of ponds and slow-moving rivers in the southern United States. Because it invades an aquatic habitat, herbicides generally are not advisable as a means of control; and because it can take over an area very quickly even if only a few plant fragments are present, removing even 95 percent of the plants with nets or harvesters is only a very temporary solution. Such species are prime candidates for biological control (see Chapter 11).

Animals are mentioned in only the first of the four factors listed above because the effectiveness of chemical and mechanical controls for widespread invasive animals depends mostly on their body size and type. Essentially, the smaller the animal, the more difficult it is to control by mechanical means. Moreover, since the impacts of chemical controls such as pesticides or poison baits are generally less specific than shooting and other mechanical controls, controlling small invasive animals is often complicated by the need to prevent impacts on native species. For example, controlling rats is much easier in an environment with no native rodents, because rodenticides can be used without causing any harm to anything but the nonnative rats. This is the strategy that is being used on some uninhabited islands off the coast of New Zealand, where rats and other introduced mammals are being eradicated with poisons in order to create safe havens for native bird species that are dying out on the mainland because of heavy predation.

## Catalyzing Long-Term Change

Complete eradication of widespread invasive species is almost always impossible with pest plants and only under certain circumstances achievable with invasive animals. Since at least a few individuals of the target species will usually still be present even after vigorous control efforts, chemical or mechanical controls should in most cases be used against widespread plant or animal invaders only when such methods provide some kind of sustainable long-term benefit. Using herbicides to kill fennel on Santa Cruz Island, for example, would be pointless if it allowed fennel to come back thicker the following year. But since fennel's abundance on the island is in part a legacy of a century and a half of overgrazing by sheep and cattle, and since that grazing pressure no longer exists, chemical control can now play an important role in a management program for the recovery of native vegetation.

Whether chemical or mechanical control of an invader will result in a self-sustaining shift toward native species depends to a large extent on what allowed the invader species to establish and spread in the first place. If a human-induced change in ecosystem dynamics is what opened the door to invasion, then controlling the invader may be a useful component of a management strategy that attempts to bring back native species by eliminating or compensating for damaging human influences on the system. For example, oak woodland habitats in the Midwest depend on fire to maintain their characteristic open structure, and invasive shrubs and trees took over the understory of oak woodlands once European settlers instituted systematic fire suppression. Mechanical and chemical control of such invaders is most effective if it is a first step in a management plan that includes the reintroduction of fire. Once the understory is cleared of brush and native grasses and wildflowers begin to take hold, a program of prescribed burning that mimics natural fire patterns will usually prevent the reestablishment of dominance by woody invaders indefinitely.

### Situations Where Continuous Control Is Justified

Sometimes mitigating the effects of human land-use practices has little or no effect on the vulnerability of an ecosystem to invasion by the target nonnative species. In such cases, it may be appropriate to use herbicides, traps, or some other form of chemical or mechanical control indefinitely if invasion would cause significant and irreversible degradation of the native ecosystem. One context in which this approach is justified is when development of a biological control program is likely and keeping the invader contained in the interim will lessen its long-term impacts. The control of paperbark trees in the Everglades is one example: Cutting and treating trees with herbicide is worth the effort in large part because several newly introduced biocontrol agents will probably keep new paperbark seedlings suppressed. Continuous control is also sometimes justified when the invaded ecosystem is endangered or otherwise irreplaceable. For example, a number of parks, wildlife refuges, and nature preserves in Hawai'i are in the process of fencing their boundaries and shooting or snaring all pigs, goats, and other large feral mammals found within their borders. The fences will need ongoing maintenance and cleared sections will have to be checked for intruders indefinitely, but the alternative is to allow the complete destruction of the native forest ecosystem these parks and preserves were established to protect (see Chapter 1).

## Selecting Appropriate Methods of Chemical and Mechanical Control

Once chemical or mechanical methods in general have been determined to be an appropriate means of controlling a given invasive species, a number of considerations need to be taken into account before selecting a specific control method. Below is a brief examination of the most effective methods of control for various kinds of species. This is intended only as a general overview; for more specific information, consult the references listed in the notes for this chapter or ask a county extension agent or some other knowledgeable source.

### Control Methods for Plants of Different Types

The most appropriate method of control for a given species of nonnative plant is strongly influenced by the natural history—that is, growth form, resprouting ability, seed production, and the like—of that particular invader. The book *Plant Invaders* offers specific information on control of a wide variety of species. What follows is an analysis of some of the trends that emerge from the specific management prescriptions offered in that book.

Trees are generally an easy target for a machete or a chainsaw. For some species, merely cutting down the existing trees may be enough to control the population for long periods. Most pines, for example, do not resprout from stumps or roots after their main stems are cut. However, cutting may release seed germination by opening up the site to light and removing the inhibitory effect of competition from established trees. Burning may then be used in some settings to eliminate such seedlings, and "cut and burn" is a common recipe for controlling invasive pines.

Other tree species resprout from the stump or root if their main stem is cut. These species must be grubbed out completely to remove the roots, or the stump must be treated with an herbicide strong enough to kill the root system. For example, Australian paperbark tree is a "cut-and-stump-treat" type of tree.

Shrubs and small trees can be treated in various ways including cut, cut and burn, cut and stump-treat with herbicide, application of herbicide directly to the bark of the trunk, injection of herbicide capsules into stems, spraying foliage of uncut plants, or grubbing out of entire plants. The larger number of stems common in shrubs and small trees makes the work of cutting more time-consuming, and herbicides or fire are often the preferred means of control.

Climbing plants of all kinds are laborious to control by cutting, as many runners must be collected and cut. Dipping cut ends in herbicide may increase control efficiency. Application of foliar herbicides to uncut plants is possible only if the vine is draped over vegetation that is also slated for control (for example, nonnative trees). Vines whose seeds sprout high in trees and send aerial roots downward are even more difficult to control, for they are already well established long before they become accessible.

For herbaceous broadleaf plants, slashing, hand pulling, applying foliar herbicides, burning, and selective grazing are options, and the method that works best for any given species often depends on the context of the plant to be controlled (that is, whether it is surrounded by sensitive native plants or by other invaders). Grass control methods include burning, applying foliar herbicides, and grubbing out (for bunchgrasses). Complete control is rarely possible for established invasive grass species, and management usually focuses on suppressing alien species enough to give native grasses a competitive advantage.

## Control Methods for Invasive Animals

Mechanical or chemical controls have been used against a variety of mammals, including goats, deer, pigs, rodents, rabbits, and cats; a few fish, particularly lampreys in the Great Lakes; and a few birds such as gulls. Other projects have attempted to eradicate or suppress damaging nonnative insects in natural areas, often without any lasting success. Differences in body size, abundance, population biology, and reproduction rates make each kind of animal a separate and unique case, and there is an extensive literature that explains the various methods and documents their effects.

Large animals such as sheep or cattle may be rounded up and removed if the terrain is not too rugged. However, some animals can be quite difficult to catch and must be killed. Shooting is a common method of control for goats, sheep, and pigs. Complete eradication of nonnative mammals is often feasible on small islands, since the control area is of a manageable size. However, if the area to be cleared is only a portion of some much larger infested area, fencing becomes a critical tool to ensure that areas cleared of unwanted vertebrates are not reinfested. For herd species, such as goats, the last few difficult-to-locate animals can be found using radio-collared "Judas animals," which lead hunters to the remaining groups. This is a critical step, because if a control operation is abandoned before every animal is killed, the population will quickly rebound to original numbers and any benefits of the control program will be lost.

Hunting is ineffective for smaller animals, not only because a small animal is a more difficult target, but also because smaller animals generally have higher population densities. Some species, such as muskrats and nutria, can be eliminated by intensive long-term trapping. When there are no native species that are closely related to the target species, poison baits can be useful for eliminating or at least suppressing species such as cats, rats, and rabbits. Snares can also be useful to help clear larger animals like pigs from fenced areas, provided there are no native mammals or birds that could be caught in the snares. Isolation of cleared zones from other infested areas is critical to preventing reinfestation.

Nonselective fish poisons such as rotenone can be used to clear small ponds, lakes, and isolated stretches of streams of all fish. This is a drastic step that can be justified in some cases to rid isolated habitats of invasive species of fish so that they may be restocked with rare or sensitive native species that need invader-free habitat if they are to recover. This approach makes sense only when the gain in habitat for the native species is more valuable than the individuals of that species that will be killed in the process. In some cases, the native species to be protected may have died out completely from the water body to be treated, making the evaluation process simpler.

Selective fish poisons are relatively uncommon. One of the few exceptions is in sea lamprey control, where a selective poison is applied to breeding streams to kill the larvae of these parasitic invasive fish without harming native fish. A program of sea lamprey suppression using this approach has been under way for many years in the tributaries draining into the North American Great Lakes to reduce impacts of sea lampreys on native fish species in the lakes.

Eradication of invasive nonnative insects has been attempted, but is rarely successful. The Mediterranean fruit fly (commonly known as the medfly) is "eradicated" from California almost every year. The economic losses to California fruit growers should this species become common would be enormous, so the U.S. Department of Agriculture and state agricultural personnel use a system of traps baited with attractive odors to identify infested areas. These infestations are then treated with a combination of pesticide-treated bait and releases of sterilized flies that mate with wild flies, causing wild females to lay infertile eggs. It is hotly debated whether the frequent need to "reeradicate" this fly in California is due to frequent reinvasion or to control techniques that suppress the flies for a year or two to levels too low to detect. In any case, the failure of agriculture officials to rid the region of medflies permanently—even with huge expenditures and massive effort—demonstrates the near impossibility of eradicating insects with chemical or mechanical controls.

A few other efforts have been made to eradicate newly recognized invasive insects or use chemical or mechanical means to protect native species from their attacks. In the late 1990s, quarantine officials cut thousands of maple trees in Brooklyn and Chicago in an attempt to prevent the establishment of the Asian longhorned beetle, an insect whose larvae infest and kill native maples and other hardwoods. The pest arrived as larvae inside the wood of packing crates used to enclose commercial goods exported from China to the United States. It is unlikely that even this "scorched tree" policy will succeed at eradicating the invader (see Chapter 6). Insects are so small and reproduce so rapidly that it is almost never possible to kill every last one.

However, chemical or mechanical control of insects can play a role in protecting threatened plants at specific locations. For example, the rare semaphore cactus may be at risk of extinction from the invasive insect *Cactoblastis cactorum*, a cactus-feeding moth from Argentina introduced into the Caribbean in the 1950s for biological control of pest cacti and later spread to Florida by the plant trade (see Chapter 11). There are only a handful of individuals of this endemic cactus species still surviving in Florida, mostly because of habitat loss. Use of pesticides or mesh cages to protect the remaining plants would be an effective temporary means of conserving the species while researchers investigate other solutions, but neither technique would have any effect on the pest population itself.

## Chemical and Mechanical Control in Action: Protecting and Restoring the Hakalau Forest in Hawai'i

The Hakalau Forest National Wildlife Refuge in Hawai'i is a 33,000-acre tract of native forest that is slowly being rehabilitated to preserve habitat for native endangered forest birds and plants (see Chapter 1). This conservation area consists of a series of forest types along an altitudinal gradient from 2,500 to 6,000 feet. From about 6,000 feet to the refuge boundary at 6,500 feet, what was once koa forest is now abandoned pastureland dominated by alien invasive grasses and shrubs. Even in the native forest that remains, a variety of invasive species threaten the integrity of the forest community and the survival of the rare and endangered species the refuge was established to protect. Among the most urgent tasks facing the refuge are restoration of native forest in deforested areas and protection of existing native forest from invasive plants and animals.

Though there are dozens of invaders of concern on refuge lands, alien species control efforts are focused on pigs, cattle, blackberry, American holly, banana poka, gorse, and cane tibouchina. These invaders were selected for immediate control because they are the most serious threats to the preservation and restoration of native habitat at Hakalau.

Habitat restoration began with fencing to control cattle and pigs. By 1999, the upper third of the refuge had been fenced into a series of 1,500- to 2,000-acre management units, and staff and professional hunters had killed nearly all the cattle. Eradication of cattle was relatively straightforward because they were few in number and were easy targets for marksmen. Eliminating pigs from the refuge has for various reasons proven much more difficult, and progress with pig control is slow.

The refuge first tried public hunting as a means of pig control, but later abandoned this method because hunting pressure was not high enough to reduce pig populations below sustainable levels. The refuge then used professional hunters and dogs to keep constant pressure on pigs in a given unit until it was cleared. In 1998, refuge staff began using snares to control pigs with great success, but as of the end of 1999, significant portions of the refuge had yet to be cleared of pigs. (See Chapter 1 for a more in-depth discussion of the mechanics and politics of pig control.)

Removal of these animals has had immediate impacts on the ecosystem. Regeneration of native plants has increased significantly, which is one of the main goals of restoration. Unfortunately, some nonnative plants have also benefitted from the release from browsing pressure, especially a blackberry species (*Rubus argutus*) imported from the southeastern United States that has invaded the midelevation forests. This plant is the number-one invasive plant threat to the refuge because it invades intact forest and is spreading rapidly now that cattle no longer suppress it.

Had the refuge staff foreseen this plant's potential for rebound following cattle elimination, the initial infestations could have been controlled as cattle were removed for a few tens of thousands of dollars. But by 1999, nine years after cattle removal, blackberry was everywhere and its control now is likely to cost $500,000 to $1 million. However, while this pest might appear to be a difficult-to-hit species—it lacks single, distinct stems and is often intermingled with desired vegetation such as young koa trees—it can be controlled with a very low concentration (0.5 percent) of the herbicide Garlon 3A®. Applied as a foliar spray, this herbicide kills the blackberry without harming native trees.

In this same forest type, refuge staff also had to exterminate American holly, the familiar and much loved Christmas plant. A section of land in what is now refuge was once used to grow this species as a crop for the Christmas market, and both native and exotic birds ate the berries and spread them into the forest. When the refuge began a control program for this invader, about 500 acres were infested with holly. However, the species has a growth form that makes it a relatively good target for chemical control. Most holly plants have single trunks that can be killed with EZJect®, an applicator lance that uses a .22 shell containing an herbicide gel that is pressed into the cambium layer of the trunk. Control work can be done by volunteers because the herbicide does not have to be handled directly, and each trunk need only be injected with one capsule per centimeter of trunk diameter. Injected plants do not resprout from the roots, so retreatment of individual plants is not necessary. However, treated areas must be scouted annually for seedlings, and foliar applications of herbicide are used to kill new plants until all the holly berries in the soil have sprouted or died. No one knows how long this will take.

Banana poka is an invasive vine from South America that grows up into the canopy of native trees on the refuge, forming large curtains that drape over and shade native plants. Heavy infestations kill native plants and prevent regrowth. Eventually, infested areas become monocultures of banana poka. Control requires prior removal of pigs and cattle, which disperse the seeds. Manual removal of vines using volunteer labor is effective in reasonably accessible areas that can be revisited periodically until the seed bank has been exhausted. The plant is now mostly under control in the accessible parts of the refuge that have been fenced and have had pigs removed.

Gorse is the fourth plant on refuge's work program. This flowering shrub is not a forest invader, but rather infests the 2,000-acre section of high-elevation abandoned pastures that the refuge wants to reforest. Because there are no native plants in these old pastures at the moment, more extreme methods can be used without risk. To kill the gorse, crews spray the bushes in spring with an herbicide that kills most of the existing plants. Any larger gorse plants in the treatment areas that are not killed by foliar treatments receive basal bark applications, in which a more concentrated dose of herbicide mixed with fuel oil is applied directly to the lower trunk. Touch-up treatments to kill new gorse plants that sprout from the seed bank are generally necessary for the first few years after native trees are planted. As native trees grow and create shady conditions, young gorse is unable to survive and the job is done. As of 1999, the original 2,000 acres infested with gorse had been reduced to only 75 acres.

Cane tibouchina is a small herbaceous plant (related to the invasive miconia tree) that has been found in the understory of open-canopy forests on the refuge. It is very difficult to control because it regrows readily from plant fragments such as pieces of stem or root. If left uncontrolled, it dominates the herbaceous layer of the forest and excludes almost all native plants. This species can be controlled only by digging out the entire plant, an extremely laborious task. Fortunately, there have so far been only a few dozen plants discovered on the refuge, and each one has been pulled up, bagged, and removed from the property. However, 33,000 acres of rough terrain offer many hiding places, and there is always the risk that populations may incubate in some out-of-the-way place until they are too extensive for refuge personnel to remove.

Have chemical and mechanical control methods restored native habitat in the Hakalau Forest? Yes and no. If the refuge crews were laid off and the fences allowed to rust, pigs and cattle would return to their original population levels within a decade or so. Holly might one day be eliminated from the refuge, but birds will continue to disperse holly seeds from sources outside the refuge into holly-free areas. Blackberry is likely to require a intense and prolonged control effort, and it, too, would be reintroduced to the refuge if control efforts ceased and pigs (one of the principal dispersers of its seeds)

Fire was used as a method of controlling gorse in abandoned pastures at Hakalau Forest until it was discovered that burning stimulated the growth of new gorse seedlings. (USFWS Jack Jeffrey)

The first step in gorse control today is to spray the plants with herbicide. (USFWS Jack Jeffrey)

Once gorse has been cleared from a site, volunteers and staff plant native koa tree seedlings to begin the process of forest regeneration. (USFWS Jack Jeffrey)

Within a few years, the koas create enough shade to prevent reinvasion by gorse.
(USFWS Jack Jeffrey)

remain in the refuge. Cane tibouchina is a time bomb that may or may not
be defused. Gorse removal is the only permanent restoration, since this
invader cannot compete once forest reclaims the pastures. People have
become a necessary ecological symbiont of the refuge, for it is only with
human assistance that the natural community can survive in its native form.

The long-term prospects for conservation of native species on the refuge
illustrate both the power and the limitations of chemical and mechanical
control methods. The land that is being restored is of high ecological value,
but only so long as people are willing to invest time and money in its protection. While the forest is in significantly better condition as a result of
these efforts than it would have been if nothing had been done, little of what
has been accomplished is self-perpetuating.

Except when they are used immediately to eradicate a new invader, chemical and mechanical controls are only a stopgap. They are often quite useful
and sometimes are essential, but only in a minority of cases do they alone
restore balance to a system that has been severely disrupted by invasive
species. Chemical and mechanical control methods are most likely to have
permanent effects in smaller areas that can be fenced or that are naturally

isolated. As areas infested become very large, these tools become unafford-able and too environmentally damaging for effective long-term use. The key to successful control of invasive species lies in combining these short-term methods of control with longer-term changes in land management or intro-duction of carefully selected biological control agents. Restoring balance is rarely straightforward, and there is often more than one right tool for the job.

## CHAPTER 10

# Fighting the Green Wildfire

## *Integrated Management of Leafy Spurge on the Great Plains*

Eastern Montana is empty country. There are no mountains and not much that could be called hills, just mile after mile of gently rolling shortgrass prairie, punctuated every so often by a draw or a fence or a road. Ranching is one of the few ways people can use this kind of land and make a living. Even though ranches are few and far between and sometimes run to hundreds of thousands of acres—for it takes a lot of land to support a cattle operation in a place this dry—the land is the foundation of the local economy. Directly or indirectly, many of the people of the northern Great Plains set their tables with the proceeds of these ranches.

The Ruggs are one such ranching family. In 1951, Glenn Rugg bought a 14,000-acre ranch outside of Plevna, Montana, and began running cattle. But the second half of the twentieth century was particularly hard on ranchers and farmers—especially on the western Plains, where even in a good year it takes at least 30 acres to support a single cow and her calf. The difference between a few bad years and bankruptcy has grown almost impossibly thin in this unforgiving part of the continent. Like millions of others, the Ruggs had to fight to keep their operation profitable through decades of falling commodity prices and ever-rising costs. Keeping the ranch viable became harder every year.

By the end of the 1980s, it looked like it was all over. Half a decade of drought and exorbitantly high interest rates had pushed Glenn Rugg's operation past the point where he could keep it afloat, and in 1989 he was forced to declare bankruptcy. All the usual culprits—falling cattle prices, not enough rain, rising costs of labor and machinery—were implicated in the

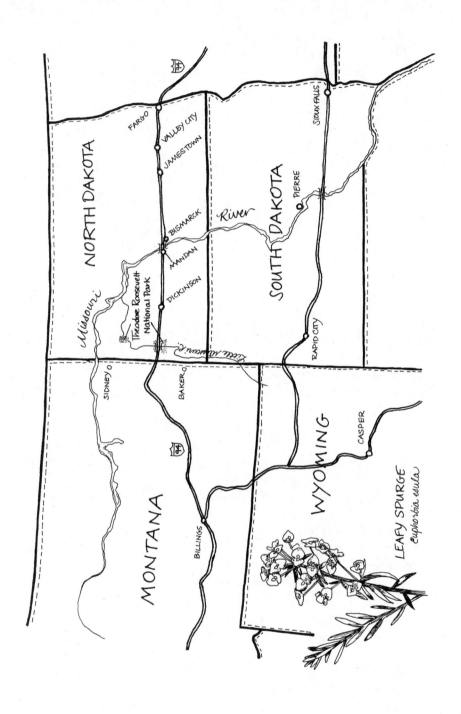

NORTH DAKOTA

SOUTH DAKOTA

MONTANA

WYOMING

FARGO

VALLEY CITY

JAMESTOWN

BISMARCK

MANDAN

DICKINSON

SIDNEY

BAKER

BILLINGS

Theodore Roosevelt National Park

Missouri

River

Little Missouri R.

SIOUX FALLS

PIERRE

RAPID CITY

CASPER

LEAFY SPURGE
Euphorbia esula

failure of the ranch. But there was another factor as well—one that is rarely mentioned in articles or books about the difficulties facing American farmers, but that in the long run would have been just as significant as lack of rain in making the ranch go into the red. The Rugg ranch was slowly being taken over by leafy spurge, an invasive nonnative weed that cattle refuse to eat. Every year, more and more prime grazing land was rendered useless by dense stands of spurge. While infestation was not quite big enough yet to be a major factor in the bankruptcy, the weed was poised to become a major drain on the ranch's productivity within a decade or so.

The Ruggs managed to reorganize and hold on to their land, but it was looking like it was only a matter of time before bankruptcy hit again. Spurge was spreading, and the Ruggs didn't have the money or the time to spray it anywhere but in their hayfields and around the ranch buildings. But in 1990, Glenn and his son Duane heard about a new control method that gave them reason to hope they might still keep the ranch. Researchers at a U.S. Department of Agriculture's Agricultural Research Service (ARS) had recently identified, tested, and released several species of tiny flea beetles that fed almost exclusively on spurge and appeared to be capable of reducing solid stands to scattered plants in less than a decade. The Ruggs contacted the ARS lab in Sidney, Montana, and arranged for release of beetles on their ranch. They had nothing to lose.

Seven years later, Duane couldn't say enough good things about the beetles and about the biological control program that had brought them to the ranch. He showed me photos of hillsides that had been entirely covered with spurge only a few years before and then pointed to the same slopes now thick with grass. Spurge was still present in patches, but he expected that within a couple of years it would be reduced to incidental levels. The family's ranch was coming back.

"What do you think this is all worth to you?" I asked Duane.

"It's worth the ranch," he said without hesitation. "There basically wasn't much hope for continuation."

Duane would be the first to acknowledge that alien invasive species are a major threat to the grasslands of the Great Plains—and that biological control directly benefits landowners like himself who depend on the biological integrity and productivity of the Plains for their livelihood. Controlling spurge has become a bottom-line issue for his family, and although herbicides are still useful in some situations, spurge would be unmanageable without the beetles.

Duane has been trying for years to get his neighbors to quit relying strictly on herbicides and sheep for spurge control, but he hasn't been nearly as successful with them as he has with spurge. "They're skeptical about it for the most part," he explained, shaking his head. "My other neighbor showed me some spurge in his draws. I tried and tried to convince him to come over and get some bugs, but he didn't have any interest." Attitudes change slowly, and getting people to adopt a new technology for handling an old problem is in many ways even harder than developing the technology itself. "I guess people have just been spraying for years and years," Duane said, "and it's just hard to try something different."

## The Making of a Weed

Leafy spurge first took root on the North American continent far from where it eventually became a major weed. The plant was initially discovered in Massachusetts in 1827, where it had probably come in as seed in ballast soil from ships sailing from Europe. It did not surface again in the literature until 1848, when it was listed in Gray's *Manual of Botany* and described as "likely to become a troublesome weed." But for several decades spurge remained an incidental component of the northeastern flora. Though it was reported in New York in 1875 and in Michigan in 1881, it remained uncommon across this initial range.

But once spurge hit the Great Plains, it came into its own. By 1910, it had crossed the Mississippi and taken hold in the rangeland of the West. The weed was probably introduced to North America multiple times, and it is hypothesized that the Great Plains infestations started with crop seed brought in by Mennonite immigrants from the Ukraine. Spurge is a tough perennial that thrives only on land that is not regularly plowed, as its strength comes in part from the massive root network it builds up over years of growth. While it eventually became a nuisance in eastern states, where it sometimes infested pastures and required expensive control, it was on the expansive but degraded grazing lands of the northern Plains states that spurge populations grew exponentially. Infested acreage in the region doubled about every ten years, and by the mid-1990s, spurge had taken over almost 5 million acres of land.

Spurge did so well on the Plains in part because much of the land there was already seriously degraded long before spurge arrived. During the early 1900s, most of the native prairie of the northern Great Plains was turned under by settlers who, under the Homestead Act, were required to plow a portion of their land in order to receive title to it. As a consequence, places where the soil should never have been plowed were turned under and farmed. When the soil began to blow away from lack of cover, the govern-

The leafy spurge plant, an invasive perennial from central Asia, has taken over millions of acres of rangeland on the northern Great Plains. (Jason Van Driesche)

ment helped farmers replant with a variety of nonnative grasses, many of which proved invasive. The combination of plowing, introduction of invasive species, and subsequent overgrazing left the Plains in such a vulnerable state that spurge was essentially offered an open door.

Once it reached this favorable habitat, spurge's explosive spread was driven largely by a single factor: the absence of its controlling natural enemies. While its large root system, toxic sap, resistance to herbicides, and copious seed production certainly made it harder to control than many other invaders, these characteristics alone did not make spurge a major weed. In its native habitat, spurge is not invasive specifically because a wide variety of specialized herbivorous insects have developed a tolerance for the toxic compounds in its sap and do enough damage to the plant to keep it in check. Since these insects did not accompany leafy spurge when its seeds were brought to the New World, spurge had a decided advantage over most native plant species on the Great Plains. In established spurge infestations, the invader forms an ever-denser stand of vegetation that nothing native to North America can eat.

The thickest and most extensive infestations were in Montana, North Dakota, South Dakota, and Wyoming—states where ranching is an important industry and rangeland a major resource. Amazingly, it wasn't until

1991 that anyone attempted to quantify the economic impacts of the spurge problem. In an analysis of the relationship between spurge infestation and rangeland value and productivity, agricultural economists at North Dakota State University found that a 40 percent leafy spurge infestation cut the herd carrying capacity of rangeland in half. Given such a reduction in capacity, they calculated that leafy spurge had cost ranchers in those four states about $33.5 million in lost income in just one year. The indirect impact on the regional economy—that is, the ripple effect of lost ranching income on retail sales, services, and other economic activity—was estimated at an additional $76.6 million, for a $110 million total annual loss. Given the rate of spread of leafy spurge, the researchers predicted that if spurge was not brought under control, total economic impacts on grazing land could reach $144 million per year by 1995—a 31 percent increase in just five years. So long as spurge kept expanding its range, it was going to keep putting ranchers out of business.

Though this economic analysis did not specifically address the ecological consequences of the spread of leafy spurge, the magnitude of its impacts on native ecosystems was made clear in other studies. Most of the area invaded by spurge is shortgrass and mixed-grass prairie, and while much of the region had already been seriously degraded by overgrazing and by other invasive plant species, native plants were still likely to have been present in many areas before spurge invaded. Wherever spurge took hold, though, populations of almost all native grasses and forbs plummeted. One study determined that leafy spurge was "clearly related to a decline in the abundance of the dominant [native] species . . . both on a large scale and within a single infestation." In spurge-infested areas, the study found that four out of five common native grasses were almost entirely absent and that two alien pasture grasses were the only other species that increased in abundance as spurge consolidated its hold. In addition, since spurge is toxic to most wildlife, the native grazers of the Great Plains probably avoid spurge-infested areas just as cattle do. While some birds are able to eat its seeds, the overall effect of a spurge invasion is to push out most remaining native species.

Rapidly increasing economic and ecological impacts have for years made spurge a high-priority target for control. But from the turn of the century to the late 1980s, spurge kept spreading despite many attempts to stop it. These years had to be frustrating for the government agencies and private individuals who could only watch as spurge took over 10 acres for every acre they knocked it back. In the long run, though, the time and money spent on developing chemical, mechanical, and cultural controls were worth the investment. Even though herbicides and sheep control alone weren't enough to stop spurge, developing these technologies was the first step in pulling together a broad program of integrated pest management (IPM) for the weed. Only such a flexible and intelligent approach to weed manage-ment could provide benefits to landowners as well as to the ecosystem, for controlling a weed as adaptable and tenacious as leafy spurge demanded a toolbox with many types of tools.

## Like Spitting on a Wildfire

The first tools for spurge control were admittedly quite crude and generally created more problems than they solved. In the early 1920s, for example, the New York Agricultural Experiment Station proposed spreading 10 tons of salt per acre on spurge or saturating the ground with kerosene around spurge infestations. The assumption behind such methods was that it was worth killing a patch of soil entirely in order to get rid of spurge, and highly toxic substances were used in large quantities in the name of eradication.

These and other "scorched earth" approaches were abandoned piecemeal as it became clear that total eradication was unlikely. From the mid-1930s to about 1950, a limited suite of chemicals—sodium chlorate, sodium arsenite, calcium cyanamid, and arsenic pentoxide—were used with little success for spurge control. Some research was done on the efficacy of goat or sheep graz-ing as a means of control, for unlike cattle, goats and sheep tolerate spurge well, but the method never caught on. Even after World War II, spurge was still regarded more as a nuisance than as a catastrophe. Control efforts through the 1950s were desultory at best—this despite the fact that, in a 1951 survey, 30 percent of North Dakota farmers "were concerned that leafy spurge was taking over their farms."

By the early 1960s, the invasion had become too massive to ignore any longer. One North Dakota researcher noted that "it was becoming obvious to any casual observer that leafy spurge had increased in acreage during the past decade." Armed with new herbicides, the North Dakota County Agri-cultural Agents Association launched a program of all-out chemical warfare in 1966 with spurge control demonstrations across the state. Similar efforts were undertaken in other states, but interest faded as it became clear that spurge was too entrenched for even improved herbicides to have any real effect on established infestations. Faced with escalating costs and no real victories, these campaigns were largely abandoned by the early 1970s. According to the report cited above, "no treatment could eradicate the weed. Individual efforts had failed or at best were holding their own in an increasingly expensive control program."

Spurge had become a wildfire, and no one on the prairie had the tools to put it out. Herbicides rarely eradicated an established patch of spurge—even the strongest chemicals only killed the foliage, leaving the roots to resprout within a year or two. Landowners who sought out new infestations and controlled them aggressively from the start stood some chance of staying on top of the problem, but most people weren't that vigilant. Compounding the problem was the fact that spurge control in most states was coordinated and funded largely at the county level. Many county governments had neither the money nor the expertise to scout for and control spurge as thoroughly as the situation demanded. Even if one county mounted an effective control campaign, its efforts would soon be for naught if a neighboring county's landowners let spurge multiply unchecked. Local solutions were no longer adequate, for spurge had become a problem of regional scope and magnitude.

Control efforts languished until the late 1970s, when the first Leafy Spurge Symposium in Bismarck, North Dakota, brought together a group of researchers, extension agents, and landowners from across the region to address the spurge epidemic. The conference generated broad support for leafy spurge research and control, and prompted the heads of several key agencies and organizations (among them the state and federal agricultural research stations of the Plains states) to submit a joint proposal for spurge

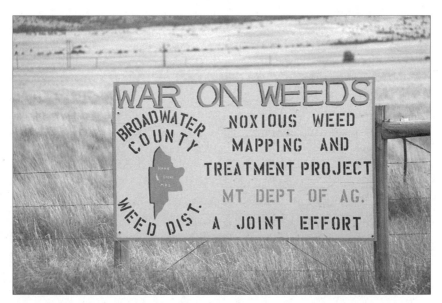

Weed control is an important function of county government in the Plains states. (Jason Van Driesche)

research to the Old West Regional Commission. These same agencies also began to redirect a substantial portion of their own resources to investigating a variety of spurge control methods, from optimal spray rates to the efficacy of sheep grazing, to basic spurge biology. The knowledge base for spurge control began to grow rapidly.

As research efforts picked up speed, state agencies began to put their own resources into developing a more sophisticated and broad-based control infrastructure. In the early 1980s, North Dakota launched an effort that not only advocated spurge control statewide, but also coordinated and (for the first time) funded control at the state level. In 1983, state and local agencies and individuals formed the North Dakota Weed Control Association to develop and implement a statewide program for spurge control. The state Department of Agriculture began working with the county weed boards, and a special tax was imposed for weed management. However, while the political base of the association was strong enough to continue to generate greater financial support for spurge control, these resources were spent almost exclusively on herbicide spraying programs. The counties were spending little or no money on planning or carrying out an integrated approach to the crisis.

Even so, it was at about this time that the scale of the response began to match the scale of the problem. Although individual counties were focused mostly on supporting chemical control, a consortium that included Agriculture Canada, the USDA's Animal and Plant Health Inspection Service and Agricultural Research Service, and the states of Montana, Wyoming, and North Dakota were all working together to evaluate a variety of insects as biological control agents. The affected states had to form a regional organizational infrastructure before they could be ready to advocate effectively for creating a biological control program for leafy spurge. The effort had to move up before it could move on.

## With Enemies Like These . . .

If weeds are a wildfire, biocontrol is the backfire that can stop a weed in its tracks. But it takes organization and training to fight fire with fire without burning down the forest. Testing and introduction of biocontrol agents is most likely to be done well if it is done by universities, government laboratories, or other large-scale organizations. Such research is typically too complex and too expensive for local or private entities to handle effectively. The risk of biological control stems from the fact that effective biocontrol agents are by definition hardy, self-spreading alien species, and it is only through careful testing that they will consistently turn out to be "beneficial invaders"—that is, invaders that do damage only to the target species they were introduced to control. This kind of precision is part of what makes good biological control such a powerful tool for the management of invasive species. Not all biocontrol introductions have been this carefully researched or precisely targeted, especially when the goals of the program were agricultural rather than ecological. But given sound goals and appropriate protocols, biocontrol can be highly reliable (see Chapter 11).

Biological control made 1997 a year for celebration on the northern Plains. After decades of work, scientists at various agricultural research stations across the Great Plains had assembled a set of biocontrol agents that clearly had the ability to suppress even the densest stands of spurge. By the mid-1990s, this multispecies control program had shown such dramatic results that it attracted national attention. As a consequence, the USDA's Agricultural Research Service (ARS) decided to showcase its success with a tour of some of the sites that illustrated the promise of the program. In June 1997, I joined a group of regional and national ARS scientists on a road trip across North Dakota and Montana to see firsthand how a half-dozen species of tiny beetles were accomplishing what all the herbicides in the world could never do.

There didn't seem to be anything particularly remarkable about our first stop at an old pasture outside of Valley City, North Dakota. But that is precisely why North Dakota State University entomologists Bob Carlson and Don Mundal brought us here. "This area right here where you see grass— well, before we started, this was all spurge," Bob explained. "This was solid yellow." He pulled out a photo of the slope in front of us that was taken ten years ago—just before the beetles were released—and held it out at arm's length. Only the contours of the land and a few scattered trees gave any evidence that it was a photo of this same place.

Leafy spurge's continentwide retreat began on this and a few other sites around the region in 1988, with the release of just eighty individuals of one of the species of tiny *Aphthona* flea beetles that have proven so effective at controlling this weed. Presented with a virtually unlimited food supply, the beetle population exploded. By 1995, researchers, extension agents, county weed supervisors, and private landowners from all over North America were collecting millions of beetles off this site every year for redistribution to other infested areas. Much of the collection and redistribution was organized by the North Dakota Biological Control Coordinating Committee, which was formed in 1990 to distribute leafy spurge control agents across the region. This site was among the first to demonstrate the promise of the beetles for controlling spurge, and in 1997 there were only scattered spurge

Don Mundal shows how the larvae of the flea beetles used to control leafy spurge infest the plant's roots. (Jason Van Driesche)

plants across the hillside and a few patches in the draws. Chances were good that even these remnants wouldn't last long.

It was not the adult beetles feeding on the foliage, however, that ultimately loosened the weed's grip on this meadow. Leafy spurge can send roots down 6 or 8 feet into the soil and resprout from the roots almost immediately even if its aboveground shoots are completely destroyed. "The adult feeding is apparently of no consequence," Bob explained. "Spurge lives mainly in its roots." To be effective, a control agent has to do significant damage to the plant's root system. And that is just what the larvae of the various species of flea beetles do—they burrow down into the soil and eat holes in the plant's roots. This breaks down spurge's defensive system and allows soil fungi and bacteria to infect the plant tissue. "It's a double-barreled attack on the plant once the insects break down that root system," Bob said. In the wake of that attack, what was once a tremendous problem on this site is now almost incidental.

But bringing a pest down to a nondamaging level is rarely as simple as finding the single, right natural enemy and setting it loose. Few biocontrol agents will spread rapidly enough on their own to control a weed, particularly for widespread and patchily distributed weed species. Moreover, variations in microsite conditions—soil moisture, aspect, and so on—can affect

Flea beetles are barely an eighth of an inch long, but when they multiply into the millions, they are capable of reducing spurge to an incidental species. (R.D. Richard, USDA-APHIS)

the usefulness of a given biocontrol agent, so it often takes a group of natural enemy species to reduce weed density across the variety of habitats in which many of the worst weeds thrive.

Finding and testing a range of candidates for biological control is the job of both the ARS and its sister agency, the Animal and Plant Health Inspection Service (APHIS). USDA laboratories in Montpellier, France, and in Rome, Italy, were responsible for most of the foreign exploration and some of the safety testing, which consisted of collecting insects that were found feeding on native populations of leafy spurge and other closely related plants in Europe and northern Asia and testing them for specificity using potted North American relatives of spurge. ARS laboratories in Sidney, Montana, and Albany, Oregon, and an APHIS facility in Bozeman, Montana, handled some of the testing and most of the initial introductions. These two agencies have worked in close cooperation with Agriculture Canada, the USDA's counterpart to the north, which initiated biocontrol work on leafy spurge in the early 1960s.

Each of the candidate biocontrol agents was further tested for host specificity to determine the range of plant species it was capable of developing on. Testing focused on the ornamental poinsettia and over a hundred native spurges, all of which are in the same genus (*Euphorbia*) as leafy spurge. Find-

ing natural enemies of leafy spurge that refused to eat poinsettias was relatively easy, because the two are somewhat distantly related. Some of the
native spurges, however, are in the same subgenus (*Esula*) as leafy spurge,
forcing scientists to select only very narrowly specialized natural enemies.
Most of the species selected exhibit a strong preference for leafy spurge.
Others will feed on some native spurge species if leafy spurge is not available,
but do not develop successfully on any of the native spurges. Given these
results, it appears that damage to native species will be minimal.

Neal Spencer is the director of the spurge program at the Sidney lab and
has overseen the release of most of the dozen or so insects that feed on
spurge. The four initial releases that took place between 1964 and 1980—
two moths, a longhorned beetle, and a gallfly—had no significant impact on
spurge populations. The first moth has been suppressed by an insect virus;
the second moth failed to establish because it is specific to a genotype of
spurge not found in North America; the longhorned beetle is slow to reproduce and attacks only very large spurge plants; and the gallfly is kept at nondamaging levels by a native insect parasite. However, it is not uncommon
for the first few introductions in a biocontrol project to have little effect
on the target pest—just because an agent performs well in the laboratory
doesn't necessarily mean it will reach effective densities in the wild. For
example, Australian scientists had to introduce several dozen natural enemies over the course of several decades before they achieved large-scale control of the invasive prickly pear cactus with the introduction of the moth
*Cactoblastis cactorum*. In the years since, scientists have developed more
effective screening mechanisms than the Australian control effort had at its
disposal, so now it generally takes fewer introductions to hit on one that
works.

It was only with the release of the first flea beetle species in 1985 that
spurge biocontrol began to show significant promise. Over the eight years
that followed, five more species of related flea beetles were tested and
released, and each has shown the ability to kill even long-established stands
of spurge. Each has a unique set of habitat requirements as well—some like
high rainfall, for instance, while others do best in dry conditions—but taken
together, they cover most of the kinds of habitat that spurge invades. The
only places where none of the natural enemies introduced thus far has established are areas with very sandy or very clayey soils. Foreign exploration and
research continues for these remaining control needs, and as the project
matures, its benefits to landowners and to the ecosystem itself will continue
to grow.

Leafy spurge is by no means the only nonnative pest that has been successfully controlled using natural enemies (see Chapter 11). In general,

One of the flea beetle species used to control leafy spurge. (R.D. Richard, USDA-APHIS)

given careful work and the right circumstances, biological control can reduce impacts of nonnative invaders on native ecosystems far more precisely and economically than any other control option. For instance, when the ash whitefly invaded California in 1988 and began killing ash trees across the state, the release of a small parasitic wasp that exclusively attacked the whitefly reduced the pest's density on ash trees by up to 97 percent. This program saved state and local governments nearly $300 million in tree replacement costs and prevented the destruction of millions of trees. Similarly, biological control of tansy ragwort (a pasture weed across much of the West) currently provides about $5 million in annual benefit to the Oregon economy alone, resulting in reduced herbicide use and recovery of native vegetation in many areas once infested by ragwort. Experience with these and other successful management programs demonstrates that money put into developing effective large-scale biological control programs for major nonnative pests is a sound investment.

However, even when an effective suite of natural enemies has been identified and introduced, several important questions may remain to be answered. If biocontrol is not a complete solution to the pest problem at hand, how can biocontrol be integrated with other control methods? What combination of methods works best on a given kind of site? And perhaps most important, what are the most effective ways to move research results from test plots to the landscape as a whole? In 1997, the ARS launched an initiative called TEAM Leafy Spurge, whose purpose was to answer these and other questions by bringing together the results of a half-century of research and extension. Now that most of the tools for spurge control had been developed and their individual effectiveness proven, the next step was to put these tools to work clearing the Great Plains of spurge.

## More Than Just Bugs

The final stop on our statewide spurge tour was Theodore Roosevelt National Park in far western North Dakota, where TEAM Leafy Spurge was holding its first strategic planning session. But the Valley City site where we started that morning was almost on the Minnesota border, so Neal Spencer and I had plenty of time to talk about the project on the drive across the state. As we discussed the goals of TEAM Leafy Spurge, Neal emphasized one point again and again: Though biocontrol is the centerpiece, this project is about a lot more than just bugs.

"The purpose of this work," Neal explained, "is to bring all the pieces together." The last few years before the launching of the TEAM Leafy Spurge project saw the publication of a number of papers that studied combinations of management techniques with the hope of finding synergistic relationships between various control technologies. Almost all the papers emphasized that no one method could achieve control across the full range of habitats that spurge invades. But while each research project addressed some particular combination of technologies—herbicides and biocontrol, for instance, or grazing and herbicides—none of them brought together all of the options into a single cohesive framework. This was one of the primary goals of the TEAM Leafy Spurge project: to develop a decision-making methodology and a comprehensive set of demonstration plots that laid out the full range of control technologies. Ultimately, the intent was to help landowners determine which combination of spurge control methods would be the most economical and effective for their land and their needs.

On the way across the state, we paid visits to a series of demonstration sites where each of these methods was being employed. The Valley City site where we initially gathered that morning was one of the showcases of the flea beetle biocontrol. The tremendous success of the project at this location clearly demonstrated that biocontrol was to be the core element of the overall spurge management effort. But it would be misleading to extrapolate from the Valley City site to the state as a whole—Neal Spencer called this location an "ideal habitat" for the beetles and emphasized that not all sites were as amenable to beetle establishment and spread. Even under these most favorable of conditions, the beetles took six to eight years to build up to a population density high enough to suppress spurge on a large scale. This combination of factors—habitats where the beetles fail to thrive and long lag time between release and control—creates gaps in the effectiveness of biocontrol of spurge that could under some circumstances prove larger than some landowners could manage. Ranchers needed other, complementary control technologies for the places and times where biocontrol is not enough.

A few hours west of Valley City, we pulled off the interstate at Mandan, North Dakota, to visit an experimental plot where the efficacy of sheep grazing for spurge control was being tested. Since sheep will eat spurge, stocking sheep alone or sheep and cattle together will sometimes keep spurge in check. However, this kind of management is not a permanent solution to the spurge problem. Grazing leaves the spurge root mass untouched, and if the pasture were restocked with cattle alone—the livestock of choice on the Great Plains—spurge would quickly take over once again. Nor is sheep grazing a useful solution for most cattle ranchers, for the sheep have to be actively herded to keep them from eating the grass as well as the spurge. But because sheep grazing does provide an immediate reduction in spurge density, it can be a useful interim measure while beetle populations build up to damaging levels—especially given recent research indicating that grazing does not interfere with establishment or growth of flea beetle populations.

Herbicides can serve a similar stopgap role, keeping spurge infestations suppressed while flea beetle populations build to where they can provide effective control. Later that same afternoon, I visited with Ray Richard, a rancher who grew hay and pasture on his family's ranch in Belfield, North Dakota. He started having problems with spurge around 1970 and immediately began treating it with Tordon®, a powerful broadleaf herbicide. It took Ray a few years to get on top of the infestation, but he's intent on phasing out herbicides in favor of biocontrol. "I don't like chemicals," he declared. "Now that I've got it under control, I'm switching to bugs." He said he was going to spread flea beetles on his land the next year and stop using herbicides as soon as the beetle population became well established.

Research at North Dakota State University suggests that he would be well-advised to keep spraying until the beetles have actually suppressed spurge, though. A recent study showed that, if herbicide applications were timed so as not to deprive adult beetles of foliage to feed on, chemical and biological control "reduced a leafy spurge infestation faster than either the biocontrol agent or herbicides alone." Though flea beetles were the most effective single tool for spurge management, biocontrol was in some cases even more powerful if it was combined with other techniques.

## Putting the Tools to Work in the Field

It took the better part of a day to drive across the state to the TEAM Leafy Spurge meeting at Theodore Roosevelt National Park. There we joined a dozen and a half scientists, range managers, extension agents, and other public officials for a strategic planning session to develop a comprehensive strategy for on-the-ground implementation of the many technologies that had emerged over the last few decades for controlling leafy spurge. The broad goals of the TEAM Leafy Spurge project were already fully outlined and were not a subject of contention—everyone agreed that integrated control was a good idea that needed to move from laboratories and demonstration plots to the larger landscape. The focus now was on how to motivate landowners to replace their economically untenable and environmentally harmful reliance on herbicides with an integrated set of biologically based control technologies. This group of scientists and practitioners wanted to do something more than publish papers. They wanted to see research translated into results.

In the two years that followed that meeting, the TEAM Leafy Spurge project did a remarkably good job of integrating research efforts across a wide range of agencies and institutions, and invested substantial resources in creating a series of demonstration sites where landowners could see firsthand the application and benefits of a variety of control technologies. In addition, it aggressively promoted the redistribution of flea beetles across spurge's range, meeting with landowners in dozens of counties and providing similar projects in other regions with beetles and technical advice. (In 1999 alone, the project was responsible for the translocation of over 20 million beetles to locations in fifty counties in seven states.) Given the thoroughness of the redistribution program and the effectiveness of the beetles at suppressing spurge infestations, it is reasonable to assume that spurge will be an incidental species across most of the area it has invaded by sometime in the first decade or two of the twenty-first century. This is a major accomplishment, and the researchers and extension agents who made it possible deserve a great deal of credit for their work.

The logical next step in the evolution of weed control is to move beyond single-species programs to the development of a community-based capacity for integrated management of weeds in general. Weed control projects need to begin to foster active landowner support and participation in management of all weeds—a practice that has grown only very slowly on the Great Plains. Such a level of private participation in weed control will require the development of a new implementation structure that complements top-down extension with organized bottom-up implementation.

Many landowners already participate in a variety of weed control efforts.

Extension is an important part of any biocontrol program. In this photo, Don Mundal explains the life cycle of leafy spurge. (Jason Van Driesche)

However, their involvement is generally limited to individual participation in programs or incentives offered through county extension offices or weed boards. This is the traditional model of weed control in the United States, and it goes back nearly as far as does concern about weeds. In 1914, A. L. Stone of the University of Wisconsin Agricultural Experiment Station wrote that "farmers cannot afford to ignore the danger from the encroachment of these various weeds; yet there is a lamentable lack of concerted action to get rid of these enemies of crop production. . . . It is time that those interested in agriculture realized the [seriousness of the] situation." The following year, Stone listed a series of measures the station had taken to improve public participation in weed control, including a new weed circular, a weed poster, a program to add weed control to the course of study in rural schools, enforcement of the seed inspection laws, a series of meetings to discuss the weed problem, and a program of technical cooperation with farmers. None of these measures proved particularly effective in large part because they were all top-down.

Seventy-five years later, over 100 million acres in the United States alone were infested with invasive plants, and weeds were expanding their range at a rate of 3 to 8 percent annually. While the introduction of new pests is in part responsible for the continued growth of the weed problem, some of the most damaging invaders have been with us for decades. We have failed to make any substantial headway largely because weed control programs rely just as heavily today on inherently ineffective top-down approaches as they did at the beginning of the twentieth century. Instead of building a locally based capacity to handle weed problems generally, the usual approach is to try to surgically insert specific control technologies for one weed after another into a more or less unwilling public. This kind of centrally controlled implementation is the weak link in the U.S. approach to weed management and is the principal reason why control of invasive alien plants lags well behind their continued spread. Now as in 1914, a capacity for and an interest in implementation among those who use the land is the critical factor determining the effectiveness of weed control.

It is important to emphasize that the underdeveloped state of bottom-up implementation of weed control does not in any way diminish the need for centrally coordinated development and dissemination of technologies for weed management, particularly of integrated, biologically based control systems. Identifying and testing biocontrol agents and researching the interactions between various control methods is a complex and expensive undertaking that only large state and federal agencies and laboratories have the expertise and the resources to do well. As such, shifting the bulk of weed control funding away from large-scale research and into local implementation would be a grave mistake. Even the best implementation network isn't worth much if it has nothing to implement.

Even so, developing better mechanisms for moving research onto the landscape is the most urgent priority for the management of invasive alien plants. Research provides solutions, but it does not solve problems. Effective regional weed management on the mosaic of private and public lands that dominate the American landscape requires an implementation network as local and as adaptable as the pests it aims to control.

## An Advocate for Every Acre

The United States is not the only country with weed problems. Many of the same Eurasian plants that cause problems in North America—tansy ragwort, St. John's wort, and various thistles, to name only a few—are invasive in Australia as well. Like the United States, Australia has a well-developed research infrastructure and is a world leader in the development of integrated biologically based management programs for nonnative pests. But unlike their counterparts in the United States, Australian weed control programs rely heavily on a network of locally directed landowner groups to design and implement on-the-ground invasive species control. While both the United States and Australia have invested substantial resources into developing biological and integrated methods for wildland weed control, Australia has also begun to work out how best to put those tools to use.

Most of these local groups in Australia are part of a rapidly growing movement called LandCare. The movement began in the mid-1980s, after it became clear that fifty years of government-led soil and water conservation programs had failed to halt the continued degradation of the Australian landscape. In 1986, the Minister for Conservation, Forests, and Lands advanced the possibility of forming a new land protection program that would be initiated and led at the community level. The first LandCare group was formed later that same year in the state of Victoria. By the end of 1986, nine more LandCare groups had sprung up in Victoria alone, each with a focus and structure particular to local concerns and needs. By 1997, there were 2,500 groups nationwide whose membership included about 30 percent of all rural Australian landholders. LandCare clearly had hit upon an approach that worked.

According to a recent study, invasive weeds are the most important issue for most LandCare groups, and weed control is the most common kind of project they take on. Various LandCare groups have been involved in biological control projects for thirteen major weeds since 1989, generally in cooperation with one or several state- or national-level research organizations. LandCare groups develop and maintain nurseries for mass rearing of biocontrol agents, distribute natural enemies to hundreds or even thousands of field sites, and monitor establishment and assess the impacts of the agents. They also provide feedback to researchers on the effectiveness of various techniques and agents, and have begun to participate in the design of new programs.

Perhaps even more important, many LandCare groups see these biocontrol efforts as only one component of a broad-based commitment to reducing the impacts of invasive species. As the author of the above study on biocontrol and LandCare noted, "Most biological control projects (especially those involving terrestrial weeds) are long-term solutions and will form one part of an overall management strategy." Because LandCare groups are built on an awareness of the interactions between land use and invasions, they tend to take a broad view of solutions to invasive species problems. As a consequence, the LandCare movement is actually implementing an approach that has only just begun to be discussed in the United States.

The LandCare movement has proven so popular in large part because it recognizes the appropriate and necessary roles of both central government and local communities. To be successful, LandCare had to be a decentralized system with centralized support, for only a simultaneous top-down and bottom-up approach can solve land-use problems as complex and as entrenched as those facing Australia. The equation works because government provides the science and logistical support, and local communities turn these tools into concrete changes in land management.

In a book on the history and philosophy of the LandCare movement, Australian author Andrew Campbell identified several key differences between the LandCare groups and the older, government-sponsored soil conservation associations. According to Campbell, these differences are in large part responsible for the movement's growth and effectiveness:

- LandCare groups usually form to tackle a broader range of issues, not just fencing off gullies or planting trees.

- LandCare groups tend to be based on neighborhoods or catchments with contiguous boundaries, rather than on peer groups of farmers with a common interest.

- The main impetus for forming LandCare groups comes from the community, although explicit support and endorsement of government remains critical.

- The momentum and ownership of the program is with the community rather than the government.

While the cultural climate of land use in the United States is in many ways very different from that in Australia, it is conceivable that some version of LandCare could serve as a mechanism for addressing nonnative species impacts (and land-use problems in general) here as well. In fact, many models of cooperative, locally based "land care" already exist in the United States, and while none has the scope or the support of the Australian LandCare movement, several have the potential to serve as implementation networks for integrated weed management. Perhaps most prominent among them is the "watershed council" movement that first took hold in the Pacific Northwest, but has since spread to many parts of the country. Watershed councils are locally based groups that promote cooperative, proactive solutions to the broad range of issues that affect watershed health. As a consequence, the activities of any given council are determined by local concerns—agricultural runoff in watersheds dominated by farming, for instance, or logging practices in forested watersheds—and a wide range of interests take part in running the council. For councils whose self-defined mandate extends beyond the health of the water system to the integrity of the entire watershed landscape, weed management is a logical project for them to take on.

In the absence of a broad-based movement like LandCare on the Great Plains, TEAM Leafy Spurge took advantage of a variety of opportunities for more informal collaboration with local communities. For example, spurge control efforts in some of the northern Plains states had already developed a sophisticated system of both official and informal flea beetle distribution networks years before the TEAM Leafy Spurge project was launched. Bob Carlson, the North Dakota State University entomologist who showed me around the Valley City site, was particularly enthusiastic about the kind of cooperation he had seen between agencies and landowners and among landowners themselves. "The most encouraging thing that I saw on my last trip out was individual landowners out on their land collecting insects on their own for redistribution," he said. This kind of informal community participation was a significant element in the success of distribution efforts. If such participation were given a formal structure and consistent support as has been done with LandCare in Australia, it would be all the more powerful a force for the control of invasive nonnative weeds.

A more recent ARS-led control project for saltcedar, a nonnative invasive shrub in the desert southwest, was built on TEAM Leafy Spurge's model and may represent the next step toward organized community involvement in weed control. According to Ray Carruthers, an ARS scientist involved in the project, the tamarisk control consortium is designed not only to control a weed, but also to build capacity for and ownership of weed management in affected communities. A variety of interest groups—including private individuals and representatives of agricultural and conservation groups—have directed the activities of the consortium from the start. This program has the potential to begin to help local communities take on the role of active, self-organized participants in weed control as a component of regional land management generally. And this is really how integrated management of nonnative species should work—by integrating not only technologies, but people as well.

Weed control in eastern Montana could be significantly more effective if landowners had access to a network of community-based land care organizations like those in Australia—organizations that facilitated communication, education, and action by and between neighbors. The Stockmen's Association is probably too narrow in scope to take on that role, as its primary focus is political and marketing issues. Ironically, so may be Montana's county weed boards. According to Duane Rugg, the weed board in his county consists mostly of people who have little practical experience with spurge, and as a result, it has promoted policies that, from the point of view of the people who have to deal with spurge on the ground, are largely unrealistic and impractical.

What landowners in Montana (and most of the rest of the United States) do not yet have is a mechanism for addressing complex land-use issues as wholes. The spurge invasion is part of a much larger land-use problem that is too big for individual families to solve alone and too deep-rooted for single-purpose government agencies to solve for them. Research into effective control methods and extension to provide landowners with access to those methods are both essential parts of the solution. But the implementation of these methods will work best if it comes from the ground up. Especially in the case of widespread pests in rural areas, landowners like the Ruggs are a major potential ally of agencies and organizations engaged in alien species control work. Professional conservationists can do only so much, and problems this big will be solved only if every acre has its advocate.

CHAPTER 11

# The Search for Balance
## *Using Biological Control to Restore*
## *Invaded Natural Areas*

Australia was about as foreign an environment as a nineteenth-century British settler could imagine. None of the familiar European animals or plants were to be found Down Under. So the colonists set about transforming their new home into something a bit more familiar and comfortable, importing species from England and around the world. From sheep to apple trees to ornamental shrubs, the settlers brought in anything that filled the stomach or pleased the eye. And as much as they were able, they left behind any "pests" that might threaten the proliferation of the chosen members of the new ecosystem they had set about creating.

Among the ornamental plants brought in were several species of prickly pear cacti from the Americas. The dry Australian outback was an ideal environment for these plants, and few Australian insects fed on prickly pear. The cacti reproduced and spread at a phenomenal rate, moving from the cities into the bush, the grasslands, and the forests. By 1920, dense, continuous thickets of 6-foot-tall cacti covered over 50 million acres of the Australian outback. Native plants were choked out and native animals went hungry. What was a harmless native plant in the Americas was an ecological disaster in Australia.

Why did prickly pear become such a problem? Was the climate more favorable in Australia than in the Americas? Were Australian plants less able to compete against invaders than plants in other parts of the world? Was the lack of interest of local insects in eating prickly pear responsible for its tremendous spread? While all of these factors may have contributed to its aggressiveness, a successful control program carried out by the Commonwealth Prickly Pear Board demonstrated that there was one critical element in the prickly pear's success: the absence in Australia of specialized natural enemies able to limit its growth.

Starting in the 1920s, the board hired entomologists to go to the native home of prickly pear in Argentina and study the insects and diseases that attacked the plant. The board reasoned these "pests" might be useful for controlling prickly pear in Australia. Some fifty species of insects eating these cacti were found in South America and subsequently shipped to Australia for testing for their effectiveness. One species, a moth called *Cactoblastis cactorum*, proved to be the key to success. The caterpillars of this moth feed on prickly pear pads, exposing the plant's tissues to destructive pathogens. The moth was released in Australia in 1926, and by 1932 cacti began to die over vast areas. Within a decade, the combined effect of insect and pathogen reduced prickly pear to low levels throughout the whole infested area. Because this moth was a specialized feeder on *Opuntia* cacti, no native Australian plants were affected, for there are no cacti native to Australia.

This project showed that biological control—the deliberate reunification of a nonnative invader with its natural enemies—can be a powerful and ecologically safe tool to restore natural areas invaded by nonnative plants. Because specialized insects and diseases are often a major constraint on the growth of plants in their native habitats, plant importers always try to import healthy stock or use roots or seeds in order to separate a plant from the insects and diseases that attack it in its native home. The prickly pear project was effective precisely because it brought in the specialized natural enemies whose absence had allowed the plant to multiply so profusely in the first place.

The same essential concept underlies the success of control projects for accidental introductions such as leafy spurge. When spurge invaded North America, it left its natural enemies behind—in large part because it made the transatlantic voyage as seed rather than as live plants. At the same time, the specialized insects and pathogens of the Great Plains were not adapted to attack spurge, and the native generalist pests that might have fed on or infected the new invader did not do enough damage to prevent its spread. When introduced plants prosper in new lands, the reason for their spread is frequently the same as for prickly pear and spurge—lack of an effective ecological counterbalance.

Introduced herbivorous insects that become invasive in a new ecosystem often thrive for similar reasons. In their native homes, most insects that feed on plants are in turn parasitized or fed upon or infected by specialized natural enemies, from parasitic wasps to predatory beetles to fungal pathogens. The nonnative insect pests that do the most damage tend to be ones that leave these natural controls behind and encounter nothing that takes their place.

Though biocontrol has been applied mostly to plant and insect pests, the same basic principles apply to the control of nearly any kind of invader, and successful biocontrol projects have been developed for many different kinds of species. However, before biocontrol can be deemed appropriate for a given invader species, it has to pass all of the following tests:

- Does the invader cause major ecological damage over a large area, or is it likely to do so in the near future?
- Does the invader have specialized natural enemies that are missing in the invaded area, and is the absence of these natural enemies a probable contributing factor in its invasiveness?
- Are the missing natural enemies specialized enough so that they will not do significant harm to populations of native species?
- Is biological control the most practical and economical tool not only for suppressing the invader, but also for restoring balance to the system it has invaded?

This chapter explores the use of biocontrol in the restoration of natural areas that have been damaged by the invasion of nonnative species. Biological control has at times been controversial, particularly when it has been applied for economic purposes alone with insufficient attention to ecological consequences. The focus in this chapter is on the ecologically sound application of biocontrol to otherwise intractable invasive species problems.

## Putting Safety First in Biological Control

The decision to introduce the moth *Cactoblastis cactorum* to Australia was ecologically sound both because prickly pear cacti were seriously affecting native communities and because there was no chance that the agents that were introduced would attack native plants. However, these important conditions are not always met. For example, the later introduction of *Cactoblastis cactorum* to Nevis, Antigua, Montserrat, and the Cayman Islands in the Caribbean from 1957 to 1970 was an economically motivated effort to control various species of prickly pear cacti (*Opuntia triacantha, O. stricta, O. dilleniii,* and *O. lindheimeri*), some of which were native species that had become pasture weeds because of overgrazing. While the biocontrol project was initially considered a welcome alternative to pesticides, it has since become clear that these introductions were neither safe nor ecologically sound.

Unlike the introduction of this insect to such places as Australia and

South Africa, where the target was a nonnative species in an ecosystem with no native cacti, biocontrol introductions in these Caribbean islands were made against a mixture of native and introduced species in a center of diversity for *Opuntia*. Introducing cactus-feeding insects into the Caribbean positioned them to spread widely via the plant trade and come into contact with many native cacti in the same genus. For example, the semaphore cactus in the Florida Keys, already endangered by loss of habitat, is now further threatened by the arrival of *Cactoblastis cactorum*.

Australia and Nevis together illustrate that biological control in itself is neither safe nor unsafe. The difference is not the tool, but the minds and hands that wield it and the policies and attitudes that direct the process. The fundamental flaw in the Caribbean introductions was that they were based solely on economic considerations and failed to consider the ecological consequences of bringing in a major new nonnative herbivore. To ensure that future biological control projects are ecologically based, careful consideration must be given not only to the rationale for carrying out the project, but also to the degree of host specificity of agents to be introduced and the composition of receiving ecological communities where the agents will be released or to which they might spread.

## Host Specificity

An essential first step in choosing agents for introduction is to make an estimate of their host specificity or the degree to which the agent limits itself to specific host plants or prey. Such predictions may be based initially on published field host records, observations by collectors of biological control agents during field surveys in the native range of the agent, or laboratory experimentation. Characterizing host specificity is not an entirely straightforward process, and the details of how to accurately estimate host ranges and other aspects of specificity have been much debated.

While procedures are well established for herbivorous insects and plant pathogens, the process is more difficult for parasitic or predacious insects, for which unresolved technical issues remain. For example, taxonomic records and natural history descriptions for most herbivorous insects are much poorer than for plants, so investigating them as possible nontarget species that could be affected by parasitic or predacious biocontrol agents can often be quite frustrating and difficult. In addition, rearing dozens or hundreds of species of native insects for laboratory host specificity tests is far more complex and expensive—often prohibitively so—than is raising a comparable number of plant species for testing.

There are three kinds of tests commonly used to estimate the host ranges of herbivorous insects that are proposed as biological control agents. The first is starvation testing, in which the proposed biological control agent is presented with one native species of plant at a time to determine whether the agent will eat and develop successfully on that plant if it has no other option. Next are choice tests, which present several potential host plants together or sequentially to estimate feeding preferences of the candidate biocontrol agent. Last is oviposition testing, which is used to determine which plants are accepted by insects for laying their eggs.

This same approach can, with some modification, be applied to testing for predacious and parasitic insect agents, and initial efforts to develop specific protocols are under way in Australia and New Zealand. However, the costs and technical difficulties mentioned above mean that laboratory tests of the host ranges of potential biological control agents of insect pests are still much more difficult (and therefore less commonly done) than tests for control agents of plants. Testing host specificity for plant pathogens, however, is less complex—native plants related to the target nonnative plant are simply exposed to the pathogen and observed for signs of infection.

In general, tests start with species closely related to the pest and then move outward to less similar species (genus, then tribe, then family) until the boundaries of the acceptable group of host plants or insects is determined. If the agent attacks and causes significant damage to any nontarget plant species, then it probably should not be used as a biocontrol agent.

While the use of testing systems adds cost and delays, such procedures provide a high level of safety for biological control programs if the proposed agent can be shown to do significant damage only to the target pest. However, the process of host range testing is affected by many variables, including the previous experience and physiological condition of the test insects (for example, starved versus satiated) and whether the test plants are presented to the insects separately, together, or sequentially. Interpretation is rarely as straightforward as a simple yes or no, and the results of host specificity tests should always be evaluated by skilled behavioral biologists. Results from poorly constructed tests may either over- or underestimate the true host range, depending on the kinds of flaws in the testing protocol.

## Composition of the Receiving Communities

The second critical ingredient in keeping biological control introductions safe is careful consideration of the composition of the biological community into which the agent is being introduced or to which the agent may spread. Success breeds imitation in biological control, but what is safe in one ecosystem may not be safe in another. To ensure that biological control agents proposed for introduction are safe in a given context, information on the host specificity of the agent must be reevaluated in light of the species composition of local communities before the agent is deliberately spread to new areas. Geographic proximity is a particularly important issue. If an introduction positions an agent for easy spread to a much larger area, the safety of species in that larger area—not just in the area of actual introduction—must be the standard for judgment.

In deliberations about the safety of proposed introductions, rare, endangered, endemic, and evolutionarily unusual species need to be given special consideration. It is standard practice to include formally recognized endangered plants that are related to a target plant in host range testing programs. Testing of rare or endangered insects would be logical as well, but poses significant difficulties because rare insects are often unstudied and, in any case, are often very difficult to obtain and propagate in large numbers for such tests. These issues are of greatest concern on oceanic islands, which are likely to have many endemic or unusual forms. For example, the introduction of parasitoids of drosophilids (a group of tiny flies) to Hawai'i would likely be disastrous, given that the thousand or so species of drosophilids native to Hawai'i are found nowhere else on earth and constitute a species radiation of great scientific interest.

## General Lessons from Past Biological Control Projects

It has been over a hundred years since the first intercontinental movement of organisms as biological control agents, and the record of past successes and failures is highly instructive (see Appendix A). In general, biocontrol works best when it takes advantage of the ecological relationships between closely associated organisms and is least effective and most dangerous when the target and agent are only loosely connected.

Generalist feeders are nearly always a bad idea. Candidate biocontrol agents with broad host ranges are very likely to attack native species and probably will not focus their depredations strongly enough on the target pest species to control it effectively. Vertebrates, predatory snails, ants, and social wasps are never suitable for use as biological control agents, and some of the most catastrophically damaging biocontrol projects—most of which had decidedly nonecological goals—have tried to make use of such species. For instance, an attempt to suppress agricultural pest snails on the South Pacific island of Moorea by introducing a generalist predatory snail called *Euglandia rosea* caused the extinction in the wild of at least seven species of native land snails in the 1970s. Similarly, Indian mongooses that were brought to Hawai'i in the nineteenth century to control rats in sugarcane fields instead took to the woods and decimated native bird populations. These and other generalist species all have either broad host ranges or flexible foraging behaviors, making them dangerous to local native species.

The safest and most effective biocontrol agents are specialized insects and plant pathogens. There are so many species of insects and plant diseases worldwide in part because many have become extremely specialized, developing close ecological relationships with the few species they attack. Given careful research, good biocontrol agents can be found for many kinds of pests among the insects and diseases that attack them in their native habitat. Such evolved specificity means that while the ecological record of biocontrol projects that take advantage of these relationships is not spotless, it is nearly so.

## When Is Biological Control the Right Option?

Even if biological control is likely to be a safe method of control for a given invasive species, the merits of using biocontrol must always be compared to the other options available. One option is to do nothing and accept the effects of the pest. Another is to suppress the pest by altering management of the invaded habitats. A final option is to use chemical or mechanical control methods.

The "do nothing" option is the most widely used, in part because most nonnative species are not harmful, or at least do too little harm to merit the effort and expense of controlling them. Many nonnative species stay at tolerable densities and "fit in" to the communities they invade.

For those nonnative species that are clearly invasive and damaging and for which there exists a social consensus that control is needed, the first question to ask is whether human actions are an important reason why the invasive species reaches damaging densities. For example, some toxic range weeds and nonnative grasses become abundant only where overgrazing reduces competition from native species. In other cases, fire suppression or clear-cut logging may promote dense populations of invaders. In such cases, altered management of the affected areas should be considered before biological control. Reduced grazing or controlled burns, for example, may not only solve the nonnative species problem, but confer other ecological benefits on the managed area as well.

When habitat management is not effective—either because the affected area is unmanaged or because management changes do not reduce pest abundance—active control remains the only option. Active methods include chemical, mechanical, and biological control. Each, however, has its peculiar advantages and limitations (see Chapter 9).

## Projects That Are a Good Fit for Biological Control

Biological control is making a contribution to the restoration of many natural communities, some of critical ecological importance. Introductions of Australian and South American insects, for example, are helping to reverse the conversion of the Florida Everglades from native wet prairie to thickets of the Australian paperbark tree and Brazilian peppertree. A South American weevil called *Cyrtobagous salviniae* has rid lakes in Australia of choking mats of the floating fern *Salvinia molesta,* a species introduced to the region through sale for use in tropical fish tanks. Insects from Mongolia will soon be helping put water back into desert riparian stream communities in the American southwest that are being dried out and smothered by saltcedar. The list of examples could be much longer.

However, biological control is not the solution to all invasive nonnative species problems. The method makes more sense for certain kinds of problems than for others. It is especially appropriate if the invaded area is large—generally many thousands of infested acres. With mechanical and chemical methods, the cost and the nontarget ecological impact of control increases directly in proportion to the size of the area to be treated, making such methods appropriate only for relatively small infestations. For example, herbicidal control of the invasive vine kudzu over the 10 million or more infested acres in the southern United States is not practical either ecologically or economically; not only would it cause extensive physical disturbance or chemical contamination, but the annual cost would be on the order of hundreds of millions of dollars. In contrast, the cost per infested acre for biological control decreases for larger and larger infested areas because introduced natural enemies increase and spread on their own.

Biological control also is useful for making chemical or mechanical control of some invaders easier and more effective. For example, finding an effective biological control agent for the invasive forest tree *Miconia calvescens* would help keep the invader suppressed in Hawai'i even if crews with machetes and herbicides don't locate and kill every single tree. A similar combination of chemical, mechanical, and biological control is being used in the campaign to clear the Everglades of Australian paperbark trees, where a weevil has been introduced to suppress new seedlings in areas that have been cleared with saws and herbicides.

In large, remote wild areas, biocontrol is sometimes the only option for controlling alien species that do not rely on human disturbance to gain a foothold. For example, the shrub *Mimosa pigra* has invaded nearly 80,000 acres of remote, unmanaged land in the Northern Territories of Australia. Because the area has not been significantly modified or disturbed by human use in historic times, changes in land management will not suppress the invader, and the lack of infrastructure in this sparsely populated region makes chemical or mechanical control difficult.

Biological control can also be the best approach for control of invasive species that are "hard to hit" with chemical or mechanical means. Species with numerous or small individuals that reproduce easily from fragments or that leave large seed banks in the soil are all difficult to control with chemicals or mechanical methods (see Chapter 9). For example, clonal plants like Japanese knotweed can't be controlled by cutting, and unless a sufficiently selective poison happens to exist, herbicides can't be used if the invader is intermingled with native species that must be protected. Similarly, invasive insects such as the hemlock woolly adelgid that infest large areas of natural forest are impossible to control with insecticides on anything but a very small scale because the ecological and economic cost of spraying the entire forest community would be too great (see Chapter 6). All such pests are excellent candidates for biological control.

## Projects That Are a Bad Fit for Biological Control

Conversely, biological control is not the ideal solution for invasive species problems in several cases. First, biocontrol generally should not be used for small infestations that show no sign of expansion. Because the start-up costs of a biocontrol project—foreign exploration, natural enemy collection, and host specificity testing—are independent of the size of the infested area to be treated, mechanical or chemical control are often more ecologically sound and more cost-effective for pests with limited ranges. Second, even if a pest has a large range, biocontrol is not appropriate if the conservation goal is only to keep specific small areas clear of the pest. On the island of Mauritius, for example, only a few hundred acres of native forest remain on the whole island, and it would not be worth the expense to develop a biological control program for the invasive species in this tiny ecosystem. A much more cost-effective approach is to keep this forest relatively free of invasive plants by hand removal or selective use of herbicides.

In other cases, invasive species may be damaging to natural communities, but still be considered economically valuable in the region where they cause damage. Nonnative pines, for example, are sometimes pests of natural habitats in South Africa and New Zealand, yet are of great economic importance for plantation forestry. For such pests, biological control of the trees themselves is not an option because its effects could not be confined to specific areas. Consequently, cutting down the pines that have invaded important natural areas or watersheds is currently the main method of control in these areas. Biological control of pine seeds, however, may play a supporting role if suitably specific seed-feeding insects can be located. Their use would be compatible with the use of pines in plantation forestry, since natural regrowth is not necessary.

## Relative Usefulness of Biocontrol for Various Kinds of Invasive Organisms

Many kinds of organisms become invasive, from land plants and insects to mammals, birds, frogs, mussels, land planaria, snails, starfish, and marine algae. More is known about what controls the populations of some of these species than of others. While Australia is just now attempting the world's first biological control project for a marine invertebrate (a starfish called *Asterias amurensis*), biological control has mostly been used against terrestrial and freshwater plants, insects, mites, and (to a very limited extent) mammals. Unfortunately, the potential of biological control for countering other kinds of invasions is very little known.

### Plants

Since the 1920s, biological control of plants has achieved permanent control of a total of forty-one invasive plant species in one or more locations. Biological control of nonnative plants was begun primarily to restore the productivity of range- or forestlands. However, many such projects also had ecological value in that areas freed of pests, such as Klamath weed in northern California and Oregon and *Opuntia* cacti in Australia and other locations (see Appendix A), often returned to a mix of plants that more closely resembled the native community, at least in lightly managed forests and grazing lands. In the last twenty years, more weed biological control projects have been started with the primary purpose of restoring the ecological integrity of natural areas. Countries making greatest use of biological weed control have been the United States, Australia, South Africa, and New Zealand, with additional projects in tropical Asia, Africa, and the South Pacific area.

### Invertebrates

Biological control has been applied extensively against mites and insects. For these groups, specialized insects such as parasitoids and predators are frequently the key to keeping populations low. Biological control as an applied science began in California in the 1880s with successful efforts to suppress alien insect pests of citrus. Since that time, several hundred species of pest insects have been successfully reduced to permanent nonpest status in one or more countries through biological control introductions of natural enemies.

Most insect biocontrol projects have been undertaken to protect economic resources, not to protect nature. However, one class of economic resources—unmanaged forests—is virtually synonymous with natural ecosystems. Projects protecting native trees in natural forests from highly damaging attacks of nonnative species clearly have benefitted a dominant component of important natural communities, even if the projects' organizers thought of the trees as lumber at risk rather than native species under threat. Appendix A lists examples of biological control projects of such forest pests, as well as several others in which biological control of a nonnative insect has clearly had importance for the protection of native species or communities.

Nonarthropod invertebrates such as snails, mussels, and even land planaria are invading new areas at unprecedented rates, but the record of biological control for such species is poor or nonexistent. For example, the zebra mussel has invaded freshwaters of North America and threatens to cause numerous extinctions of native pearly mussels (see Chapter 4). No potential biocontrol agents have been identified. Scotland and Ireland have been invaded by the New Zealand predacious land planarian *Artiposthis triangulata*, an obligate predator of earthworms. In some invaded areas, worm populations in soils have been reduced to nondetectable levels. Near-total loss of worms may cause fundamental changes in soils and soil communities, but biocontrol of planaria has never been attempted.

In general, predators are probably too ecologically risky for use against nonarthropod invertebrates, since the fish and snails that feed on most such organisms tend to be generalists. However, pathogens or specialized parasites would be suitable agents and might be discovered with more extensive study of these pests in their areas of origin. Such studies should be undertaken; otherwise, the ecological damage from these invaders may be uncorrectable, as other means of control—poisons or hand removal, for example—generally are not feasible.

Relatively little is known about what governs the density of marine invertebrates. In temperate marine systems, algae and invertebrates are often limited by competition or generalist herbivores or predators, none of which make for safe and effective biological control agents. One exception are the sea slugs, a group of organisms that tend to feed on specific species or groups of species of sponges and other marine organisms. Some parasitic species of barnacles can be relatively specialized as well. One species—a barnacle called *Sacculina carcini*—was evaluated in the late 1990s for use against the European green crab, a species that has invaded marine waters along the West Coast of North America and may be reducing native crab populations through competition. However, while the barnacle's host range in theory did not include the crab families to which nearly all North American Pacific crabs belong, laboratory tests showed it to be lethal to several important native crabs. Biological control of marine invertebrates is a new field, and researchers will have to proceed carefully in order to develop a solid understanding of what might work for invaders of this kind.

## Vertebrates

Few alien species groups cause greater ecological damage than vertebrates. Introduced goats, rabbits, rats, cats, fish, and snakes have driven hundreds of species of plants, birds, reptiles, and fish to extinction. However, biological control has only rarely been successful against vertebrates, in large part because vertebrate populations generally are controlled in their native environments more by generalized competition and predation than by specialized natural enemies. Anything big enough to kill and eat an introduced rabbit, for instance, will in most cases prey upon a wide variety of native species in an invaded environment as well. The one exception is pathogens, and almost all successful vertebrate biocontrol has been done using specialized diseases. In Australia, the introduction of the myxomatosis virus in the 1950s reduced an enormous infestation of European rabbits by about 90 percent. Within a few decades, though, the rabbit population evolved a partial resistance to the disease and rebounded to half their original numbers. A calicivirus introduced in the 1990s has decimated the population again, but rabbit numbers are likely to increase once more as the genetic makeup of the population shifts toward resistance. This is the principal liability of vertebrate biocontrol with pathogens—while a successful plant biocontrol project is essentially a permanent solution, animal pathogens tend to lose effectiveness over time as resistance builds.

The only other vertebrate biocontrol project to date was undertaken to protect oceanic birds nesting on Marion Island off South Africa. Nearly 450,000 petrels and other seabirds were being killed annually on the uninhabited island by a group of over 3,000 feral house cats that were the descendents of pet cats left behind by a lighthouse keeper when the lighthouse was closed. A feline panleucopaenia virus was used in combination with shooting to suppress the cats, and seabird populations have begun to rebound.

Other pathogens of vertebrates have been suggested as potentially effective. A venereal protozoan, *Trichomonas foetus*, is known to attack goats and reduce fertility of infected females. Conservationists have suggested its use to slow population growth rates of feral goat populations on uninhabited oceanic islands. Use of such pathogens makes good ecological sense and would likely be acceptable in remote locations with no human inhabitants who might regard goats as economic resources rather than pests.

## Steps in a Well-Designed Program

Biological control through natural enemy introduction to suppress invasive alien species is a well-understood process that proceeds through a series of steps, each of which is important for a successful and ecologically safe outcome. With the exception of Australia and New Zealand, however, few countries have explicit laws regulating the process. An international code of good practice in biocontrol has been developed by the Food and Agriculture Organization (FAO) of the United Nations for use by countries unfamiliar with the process, but most countries do not follow the code in its entirety. The following steps are more or less analogous to the steps outlined by the FAO in that they describe every possible step, not all of which will be applicable to a given project.

STEP 1: DECIDE IF A SPECIES IS SUITABLE AS A TARGET. In the context of nature conservation, suitable target species for suppression by biological control are nonnative species that cause significant damage to native species or natural communities. Social consensus on the appropriateness of control is also essential, for some interests may perceive pest species as valuable resources. Public discussion of intended projects before they are initiated can smooth implementation considerably.

Species that are invasive and damaging might not be clearly nonnative in origin. *Phragmites* reeds, for example, are invasive in wetlands in the northeastern United States, but whether this plant should be considered alien or native is unclear because there is no specific record of its arrival in North America. As part of a proposed biological control project, DNA fingerprinting of populations of the reed in North America and Europe is under way to determine whether the population in North America is native, alien, or a mixture of the two.

Because biological control projects involve significant expense, much of which occurs in the early phases of a project, only serious pests should be selected. Species causing questionable damage over large areas or species whose impacts are limited to small areas should be either ignored or managed with other methods. However, "serious" and "large" are relative terms. An invasive species might not yet be widespread, but might obviously be spreading. Or an infested area—for example, the Everglades—might not be large in a national sense but might still be of great value. In the end, resource managers' opinions will be the best advice available on species against which biological control should be targeted.

STEP 2. DEFINE THE PEST'S IDENTITY AND PROBABLE NATIVE RANGE. The key to success in a biological control project is finding effective, specialized natural enemies that are missing in the invaded area. These species are most likely to be found in the region where the pest evolved. However, this may not be the place from which the pest came most recently. Once a pest starts to spread, it may hop rapidly from place to place in a chain of interconnected invasions. While there is no sure means to identify a pest species' native range, informed guesses can be made by looking at its distribution as well as the distributions of other species in the same genus as the pest.

When a floating fern (later to be described as *Salvinia molesta*) was discovered to be a pest in Australia, it was at first identified as a closely related species, *Salvinia auriculata*, whose range extends widely over Brazil and parts of the Caribbean. Only later was it discovered that the invader was an undescribed species with a similar range. Natural enemies that attack *Salvinia auriculata* were initially introduced but failed to control the pest. Once the pest was recognized as a new species and the natural enemies associated with it were collected and introduced, biological control proved to be a highly effective means of reducing the invader's density to nondamaging levels. Experience has shown that searches that take place on the wrong species or outside of the pest's native range are much less likely to locate effective natural enemies.

STEP 3. FIND CANDIDATE NATURAL ENEMIES IN THE NATIVE RANGE AND MOVE THEM INTO QUARANTINE CULTURES. Natural enemies may be hard to find even once the native range of the pest has been located. This stage of work takes lots of careful field work and practical travel savvy—and often a little luck. Once natural enemies are found and collected, they have to be sent by air cargo to a quarantine laboratory, where a population can be kept contained for study. Quarantine laboratories are secure buildings that employ physical design features such as sealed construction and air filters as well as strict containment protocols to ensure that nothing moves from the foreign collector's box to the outside world unintentionally.

STEP 4. CONFIRM IDENTITY, BEHAVIOR, AND HOST RANGE OF CANDIDATE SPECIES BEFORE RELEASE. Once natural enemies have been found and shipped to a quarantine laboratory, they must be separated from other material and reared in pure culture in quarantine. At this point, a taxonomist's help is needed to confirm the identity of the natural enemy. Also needed is a pathologist to certify that the natural enemy is free from obvious infectious agents. Making sure a candidate natural enemy is not parasitized or infected with pathogens is important because if agents that are released carry such organisms with them to the field, the biological control agent will itself be controlled by its own natural enemies and will probably fail to reach densities capable of controlling the target pest.

As described earlier, estimating host specificity is an essential step in the process of evaluation. Such host screening is always done for weed biological control agents and is gradually being phased in for parasitoids and predators of insects. Host range tests are often run in quarantine labs in the importing country, but field studies in the agent's native range can also contribute important information about the host ranges of candidate agents.

STEP 5. CONDUCT RISK-BENEFIT ANALYSIS OF PROPOSED RELEASE. Based on an understanding of the basic biology of the species proposed for release, its host range, and the composition of the community into which it will be released, a risk-benefit analysis must be made as a basis for deciding whether the agent is suitable for release from quarantine. Two factors must be considered.

First, one must have some sense of the consequences of inaction. Will failure to control the target pest lead to significant ecological degradation through the loss of a key species? If, for example, biological control of hemlock woolly adelgid is not attempted, eastern hemlocks and the special community they create will largely be destroyed (see Chapter 6).

Second, is the agent to be introduced safe enough? Central to this question is an evaluation of what nontarget species might be affected and whether the effects are likely to be mild or drastic. These nontarget effects must again be weighed against the consequences of doing nothing. Ultimately, one has to decide which course of action is likely to be more protective of the native system.

STEP 6. RELEASE CANDIDATE AGENTS IN TARGET AREAS. Once agents are approved for release from quarantine, biologists use laboratory cultures of the agent to rear the species and release it in suitable habitats until it becomes established, or until it becomes clear that it is not able to do so. Unless the agent spreads rapidly on its own, releases may have to be made throughout the entire region infested by the pest. Of the releases made to date, about 35 percent of agents released to control insect pests and about 70 percent of agents released to control plants have become established.

STEP 7. DETERMINE IF RELEASED AGENTS SUPPRESS THE TARGET NONNATIVE PEST. As a set of agents establish over a large area, studies are needed to determine whether the invasive species' numbers are actually reduced by the biological control program. Historically, about 16 percent of all projects have resulted in complete control and another 42 percent have provided partial control. Thus in about 58 percent of all projects directed at insect pests, the target invader's numbers have been reduced to some degree. The success rate for plants has typically been somewhat lower. Of 174 projects directed against weeds in one or more locations that were analyzed in a 1980 study, 68 projects (or 39 percent) led to some degree of suppression of the weed.

Success in biological control depends more than anything on consistent funding and persistence. The control percentages cited above are diluted by a large number of half-hearted, underfunded projects that were abandoned before researchers thoroughly investigated all possible control options or that failed to put enough money into rearing and releasing natural enemies to ensure establishment. Among projects where funding was solid and reliable, the success rate is probably much higher.

STEP 8. CONDUCT ONGOING PUBLIC EDUCATION PROGRAMS. As projects unfold, the public must be made aware of the goals and progress of the control effort. For example, educating the public about the importance of the Everglades and the mechanics of ecosystem restoration is a major component of the biological control project to suppress paperbark trees there. Only with this sort of understanding will people both understand the benefits and recognize the limitations of biological control.

STEP 9. DETERMINE IF NATIVE SPECIES RECOVER AFTER THE INVADER IS SUPPRESSED. Suppressing the invader is not the ultimate goal. Rather, the important issue is whether the native species and communities that were degraded show recovery. In some cases, this happens automatically; in others, more complex interactions may be precipitated by suppression of an invader. Removal of a grazing mammal, for example, might trigger outbreaks of nonnative plants (see Chapter 8). Similarly, biological control of one nonnative plant might be followed by increases in other nonnative plants present but previously limited by competition. In such cases, separate biological control projects may be required for two or three invasive species before a stable native plant community can be restored.

STEP 10. EVALUATE NONTARGET IMPACTS OF THE BIOCONTROL AGENT. Many biocontrol agents are considered to be effective and safe even if they occasionally attack native species, provided that the impacts on native species are not harmful at the population level. But no host range testing methodology is 100 percent foolproof, and even very carefully evaluated agents might sometimes have unforeseen impacts. Much more commonly, though, the projects that turn out to be damaging are those undertaken with non-ecological goals in mind, which cause harm to native species because they ignore or downplay ecological impacts from the start. In either case, evaluation and monitoring for nontarget impacts is an essential part of any biocontrol project because they help refine the selection process for future candidate agents. Unfortunately, this component of the process is very seldom completed, since most projects are underfunded and stop once there is a perception that the "problem" has been "solved."

## A Chronology of Invasion and Restoration: Biocontrol of Purple Loosestrife in North American Wetlands

Every biological control project is different, and not every project will include every one of the steps listed above. However, a well-considered control effort for a major pest will at least have to take each of the steps into consideration, even if it leaves some of them uncompleted. What follows is a chronological history of the invasion and control of purple loosestrife, a European plant that over the course of about two hundred years came to dominate wetlands across much of North America.

Sometime around 1800, seeds of purple loosestrife were transported in soil ballast on sailing ships from Europe to North America were deposited at ports in New Jersey, New York, and New England. Between 1800 and 1850, the weed became established over a limited coastal area in the eastern United States. It spread slowly, moving mostly with water currents and floods from one wetland area to the next. At least to a degree, it probably was spread by gardeners as well, for its beautiful purple flowers no doubt attracted attention.

A dense stand of nonnative purple loosestrife can choke out almost all native wetland plants. (Bernd Blossey)

By 1900, the weed had reached the Chicago area by spreading along canal and lake transportation systems. It reached the Mississippi River a few decades later. By 1940, the infested area covered most of the northeastern and north-central United States, and long-distance transport had created a second area of infestation in Washington State.

STEP 1: DECIDE IF THE SPECIES IS SUITABLE AS A TARGET. In the 1950s and 1960s, biologists began to recognize that loosestrife was detrimental to wildlife dependent on the cattail marshes that it invades. By the 1970s, several projects were under way to quantify its impacts on specific native species. A little over a decade later, Dan Thompson of the U.S. Fish and Wildlife Service published a study of the effects of the weed on marsh plants, ducks, and muskrats at Montezuma National Wildlife Refuge in the Finger Lakes Region of upstate New York and in other areas across the United States. His 1987 report showed that when loosestrife eliminated native cattail habitat, populations of muskrats and other wetland species declined markedly.

STEP 2. DEFINE THE INVADER'S PROBABLE NATIVE RANGE. Because the ecology of purple loosestrife was already relatively well studied, there was no doubt as to its home range: western and central Europe. Consequently, the U.S. Fish and Wildlife Service and the USDA went ahead and provided initial funding in 1985 for a biological control project against purple loosestrife, beginning with surveys in western and central Europe to develop a list of herbivorous insects associated with the plant in its native range.

STEP 3. FIND CANDIDATE NATURAL ENEMIES IN THE NATIVE RANGE AND MOVE THEM INTO QUARANTINE CULTURES. From 1986 to 1992, scientists at the International Institute of Biological Control near London organized surveys to collect and identify insects attacking purple loosestrife in Europe. The on-the-ground work was done by Bernd Blossey, a student at Kiel University, who was hired to search for natural enemies in France, Switzerland, Germany, Denmark, Sweden, and Austria. Based on both this survey and previous work, Blossey developed a list of 140 species of herbivorous insects associated with purple loosestrife.

STEP 4. CONFIRM IDENTITY, BIOLOGY, AND HOST RANGE OF CANDIDATE SPECIES BEFORE RELEASE. Between 1986 and 1992, 6 candidate natural enemies—2 leaf-feeding beetles, 1 seed-feeding and 1 flower-feeding beetle, 1 gall maker, and 1 root-boring weevil—were selected from the 140 species found on purple loosestrife for host range testing. Many of the species found on loosestrife were either too uncommonly found on the plant or too general in their feeding habits to qualify as effective and safe natural enemies, and the six that were selected were those with the most promising combination of selectivity and significant impacts. Researchers in Europe then grew specimens of North American plants related to purple loosestrife in pots and ran host specificity tests for each of the 6 candidate natural enemies. During testing, 5 of the 6 were found to have narrow enough host ranges to be considered for introduction.

STEP 5. CONDUCT RISK-BENEFIT ANALYSIS OF PROPOSED RELEASE. In 1990, the federal Technical Advisory Group (TAG) reviewed the results of host range tests in order to evaluate the safety of the agents proposed for release. (This group consists of scientists experienced in biocontrol who use an established protocol to evaluate release proposals for biocontrol of plants.) Host specificity tests suggested that some of the species selected for release might occasionally feed on two native North American plants, winged loosestrife and swamp willow, both in the same family as purple loosestrife.

Host range testing is often done on potted plants in greenhouses in order to test the widest possible range of species as quickly as possible. (Bernd Blossey)

However, tests implied that while individual native plants might suffer dam-age, the agents would have no substantial impacts on these native species at the population level. Given the serious negative effects of large-scale purple loosestrife infestations on entire native wetland communities, the TAG approved the five most specific agents for release.

STEP 6. RELEASE CANDIDATE AGENTS IN TARGET AREAS. Bernd Blossey, now at Cornell University, did the first experimental releases of the root-mining weevil in 1991 in field cages over purple loosestrife patches in the Tonawanda Wildlife Management Area in western New York. In 1992, Blossey's research team released the two species of leaf-feeding beetles at the same site. The fourth agent—a flower-feeding beetle—was released there in 1994. Unfortu-nately, the fifth agent was found to be parasitized by a nematode and was not released because no uncontaminated population could be found or produced. Releasing a parasitized control agent would have been useless and a waste of time, because the nematode probably would have kept the agent's population too small for it to have any real impact on loosestrife populations.

Entomologist Bernd Blossey of Cornell University at the site of an early field trial of the insects used for biocontrol of loosestrife. (Anonymous, Cornell Media Service)

STEP 7. DETERMINE IF RELEASED AGENTS SUPPRESS THE TARGET NONNATIVE PEST. From 1992 to 1995, Blossey made field releases at research sites across western New York to evaluate the impacts of the biocontrol agents on purple loosestrife stand density and biomass. Initial results indicated that defoliation by leaf-feeding beetles can dramatically reduce weed stands and allow recolonization of an area by cattails. Blossey began developing mass rearing techniques in 1993, and by 1995 he had enough beetles to both conduct releases at a much larger number of sites and provide other institutions and agencies with enough beetles to establish rearing sites across the country. By 1999, the beetles had become so obviously successful that beetles had been released in thirty-four states and several Canadian provinces. Research on the impacts of control agents on the weed was ongoing, using a standardized monitoring protocol to track progress of the project nationwide.

STEP 8. CONDUCT ONGOING PUBLIC EDUCATION PROGRAMS. Scientists, conservation professionals, and others associated with the program continue to talk to many groups and write articles for popular magazines that reach broad segments of the public. Others have developed special packages for classroom use, and a number of school groups have gotten involved in conducting releases and monitoring impacts.

A stand of loosestrife that was successfully controlled by biocontrol agents. (Bernd Blossey)

STEP 9. DETERMINE WHETHER NATIVE SPECIES RECOVER AFTER THE INVADER IS SUPPRESSED. In conjunction with the evaluation studies of the impact of released agents on purple loosestrife, Blossey and others are conducting research on vegetation changes at release sites to monitor recovery of native plants as purple loosestrife densities decline.

STEP 10. EVALUATE NONTARGET IMPACTS OF THE BIOCONTROL AGENT. Blossey and others are monitoring beetle behavior at release sites to determine if they have any significant impacts on nontarget species. As of the end of 1999, the only documented case where the beetles fed on anything other than loosestrife occurred when several hundred thousand adult beetles emerged in the spring and found no loosestrife. They briefly attacked a stand of multiflora rose (another nonnative invasive species) before moving on in search of loosestrife.

In the 1940s, wildfire was considered antithetical to conservation, and Smokey Bear convinced generations of Americans that it was their responsibility to keep forests from burning. Now fire is an essential management tool in many ecosystems. In the 1960s, the environmental movement highlighted the dangers of pesticides and fueled public suspicion of chemicals in general. Now some herbicides are considered indispensable tools for the management of certain kinds of nonnative plant infestations. In the 1990s, biological control came under fire as society began to recognize the possible consequences of careless introductions for native species. But once its appropriate and safe use is more widely understood, biocontrol will come to be seen as a tool that is in certain contexts essential for the management of damaging invaders.

Public reservations about biological control are often rooted in the fact that introductions are irreversible, a fact that leaves little room for mistakes. Other objections are made on principle: Since the problem is invasive nonnative species and since successful biological control agents are themselves invasive nonnative species, biocontrol introductions must therefore be ecologically "wrong." This kind of public scrutiny of biological control has in recent years served to sharpen its focus on native species preservation, and mistakes like the introduction of the *Cactoblastis cactorum* moth to North America and the Caribbean are much less likely to be made in the future.

Central to its responsible use is a commitment to implementing biocontrol in an ecologically sound manner. Good biocontrol is an open, public process that rests on both sound science and community input into decision making. It avoids the use of types of natural enemies that have caused problems in the past, focusing instead on specialized natural enemies and diseases. It conducts careful studies of release organisms in order to predict with confidence what they will eat or attack and compares that to the species found in the target regions. Finally, good biocontrol always evaluates its impacts, both on the target nonnative species and on native species in the same environment. Done with care, biological control can serve as a well-calibrated counterbalance to the destructive force of nature out of place.

# PART THREE

# Going Local in the Global Age

*What People Can Do about Biological Invasions*

# CHAPTER 12

# The Gift of Meaning

## Public Involvement in Ecological Restoration in Madison, Wisconsin

We worked our way through the woods in small groups, following transects that ran from the road down to the lakeshore. Tape measure and tree key in hand, we recorded species name and diameter at breast height for every tree over 2 inches in diameter in a series of randomly selected plots. We made estimates of canopy density and took notes on understory composition. By the end of the day, our Vegetation of Wisconsin class had assembled a mountain of data on this small patch of woods at the western end of the Campus Natural Areas on the University of Wisconsin's Madison campus. Unfortunately, none of the numbers made much sense.

Only after we compiled the data, compared them with information from other stands, and analyzed the differences between them did patterns begin to emerge. The canopy layer of this patch of woods was heavily dominated by large red oaks averaging almost 30 inches in diameter. The understory, however, was composed almost entirely of other species. Box elder, green ash, cherry, and American elm dominated the sapling layer, while young red oaks were almost nowhere to be found. What our research revealed was a community in transition.

Over the course of the semester we examined many Wisconsin community types. We tramped across wet prairies and got lost in a boreal forest and discussed the retreat of the glaciers in the middle of a peat bog. Though the patterns of change varied from one community type to another, one principle emerged that seemed to hold true across all systems. If you know how to look closely enough, clues to both the history and future of a natural community are to be found in an examination of its present condition.

247

# MADISON, WISCONSIN

Lake Mendota

FRAUTSCHI POINT

PICNIC POINT

CAMPUS NATURAL AREAS

Lake Wingra

DUDGEON-MONROE NEIGHBORHOOD

WINGRA OAK SAVANNA

ARBORETUM

Lake Monona

NATURAL AREAS

Lake Superior

MN

MI

WISCONSIN

TWIN CITIES

GREEN BAY

MADISON

MILWAUKEE

Lake Michigan

IA

CHICAGO

BUR OAK
*Quercus macrocarpa*

I have revisited this and other wooded areas around the city of Madison many times over the twelve months since the class ended, and the patterns of change that I see are troubling. The ecological heritage of many of these urban forest fragments is rapidly disappearing, and their future looks increasingly impoverished and grim. Huge old oaks that can only reproduce in an environment that burns periodically are being replaced by short-lived tree species that will never grow as old and as grand as an oak. Some stands have nothing but dense thickets of invasive shrubs beneath them, foreshadowing a day when parts of what is now woods will be little more than a self-perpetuating tangle of overgrown weeds.

Invasive species are in fact one of the most powerful forces driving a wedge between the present and the past in these urban ecosystems. Species such as buckthorn and honeysuckle, the two most common invasive alien shrubs in Madison's natural areas, disrupt natural processes by changing the structure of the communities they invade. Together with fire suppression, fragmentation, and the other effects of an urban landscape context, these invasive species are forcing a sharp break with the ecological patterns and processes that have sustained these systems for millennia. We have cut these places off from their past and set them adrift.

But the break with the past that has come to characterize Madison's natural areas has its roots in a broader ecological amnesia that humans have imposed on the entire urban landscape. Though gardening with native species is growing in popularity, most yards and gardens in the city are planted almost exclusively with ornamental species whose origins are anywhere but here and whose ecological histories have little or nothing to do with each other. What's more, many of these ornamentals are the same species that have caused such tremendous problems in Madison's natural areas. So not only do most gardens lack any sense of connection to their ecological past—they are in many cases actively destroying the future as well.

There is reason for hope, though. Savanna and woodland and prairie and woods might still be rescued from honeysuckle and buckthorn. It will take more than just restoration of trees and prairie grasses, though, to bring these places back. It will take a restoration of memory.

## Garden of Dreams

On the cover of the nursery catalog that my neighbor gave me are sprays of flowers of all colors and shapes. The heading on the first page reads simply, "Garden of Dreams." These three words capture as well as any the scope and intent of the 294-page compendium of plants that follows. Even a truly imaginative gardener would be hard-pressed to dream up a garden as wildly diverse as this one. Three thousand different trees, vines, perennials, shrubs, and grasses fill the catalog, many of them rare cultivars or species offered nowhere else. Though a scattering of more familiar species are to be found throughout, most of the plants in this catalog are fascinating, unusual, strikingly different—in a word, exotic.

Many of these hard-to-find plants are the fruits of the nursery's botanizing expeditions to distant continents. In 1998, the nursery sponsored a trip to southern Chile and to the Yunnan and Sichuan Provinces of China, making over 500 collections for introduction into cultivation in the United States. Intermittently throughout the catalog are accounts of eight other trips to various parts of Asia and Mexico that the nursery led over the preceding five years. These "plant hunting" expeditions are—for the nursery's owners, at least—clearly driven by an abiding attraction to the richness and beauty of the natural world. "We search for plants because we love studying our planet," reads the introduction to the catalog. And so they bring the planet back home to their garden in the Pacific Northwest.

The nursery's owners seem to be aware of the potential for damaging invasion that comes with importation of nonnative species. "We as gardeners, horticulturists, and botanists must appreciate both the sacredness of our own ecosystems as well as those from which seed is collected," the catalog declares. It further explains that every plant is tested "within reason" for invasiveness before it is offered for sale, and troublesome species are excluded from the catalog. In the plant descriptions that follow, though, there are numerous species that bear a warning that reads something like, "Cannot be shipped to California." These are species that have been banned by various states because they are invasive or related to species known to be invasive. It appears that awareness about invasiveness does not necessarily translate into a willingness to forgo selling invasive species.

What's more, these warnings do not mean that a gardener in California *can't* order a *Vitis thunbergii* from this catalog, or from any other nursery catalog for that matter. It just means he isn't *supposed* to. "We will send *any plant* to *anyone* in *any zone*," states the catalog in its instructions for ordering. "You are responsible for knowing the agricultural restrictions for shipping particular plants to your state." Like most horticultural companies, this nursery is selling biological explosives to anyone who wants to place an order—no questions asked.

And those species that bear warnings represent only a portion of the threat that plant importers of this sort pose to native ecosystems, for such warnings generally are applied only to species that are already known to cause serious problems. Many introduced species that eventually prove invasive have a long latency period during which they appear to be innocuous, only to spring up seemingly out of nowhere a decade or more after first introduction. For example, Japanese maple is a commonly used landscaping tree that for decades appeared to be a species that did not spread out of cultivation. A few years ago, though, I saw a carpet of young Japanese maples moving into the forest from a parent tree that had been planted at the edge of a lawn. Whether this particular species will prove aggressive enough to cause serious problems remains to be seen, for only a minority of naturalized species become pests. But if these seedlings grow and flourish at the expense of the native community into which they have been inserted, Japanese maple will join a growing list of species that have jumped the fence and gone wild.

The odds are that among the three thousand species in the catalog, there are at least a couple dozen that will within a few decades prove to be major pests. They will be the next invaders that smother the ground layer in forests across the northeast, or choke out wet prairie remnants in the upper Midwest, or fuel ever-hotter fires in the remaining scraps of Hawaiian dry forest. It is impossible to say which of the many new arrivals will prove to be the next generation's weeds. A seasoned botanist might have an idea of which families and genera are likely to contain aggressive, adaptable species, but even an expert can only make general predictions. If I had to guess, I'd put my money on those that are described in the catalog as "a lovely and colonizing species" or "renowned for its fast rate of establishment," but no one can pronounce with any certainty on the propensities of a particular species. With most invasive plants, that knowledge comes too late.

This pursuit of the exotic has been the ongoing story of ornamental horticulture. In the excitement of discovering as-yet-uncultivated plants, every generation of gardeners and botanizers somehow manages to overlook the fact that some portion of what the previous generation brought in proved enormously troublesome. This fact—that a small percentage of introduced plants will eventually become vexatious pests—seems to have little effect on the enthusiasm of gardeners and horticulturists for the next new thing.

This love of the unusual is probably a universal human trait, but it was not until the Renaissance that its expression through the cultivation of foreign plants grew from an occasional curiosity to an all-consuming passion. Modern botanical gardens were an Italian invention, for it was in Italy that the development of an effective system of global transportation first coincided with the emergence of a leisure class, making possible the importation and cultivation of exotic plants for purposes other than food or medicine. Interest in ornamental gardens spread very quickly among the continent's elite, and by the early 1600s, collectors were paying the equivalent of tens of thousands of dollars for the latest and the showiest flowers and trees and shrubs. Gardens had become a status symbol for an increasingly wealthy and urbane upper class. "I do believe I may affirm," declared one English writer of the late seventeenth century, "that there is now in 1691 ten times as much gardening around London as there was in 1660." No doubt most of the enthusiasm was fueled by the introduction of alien plants by explorers of distant continents.

A significant portion of these early collections were made in North America by naturalist-explorers who sent hundreds of native North American species back to Europe as botanical samples or for introduction into cultivation, among them such familiar species as Fraser magnolia and Douglas fir (both named for the explorers who introduced them to Europe). There was little deliberate movement of ornamental species from Europe to North America during colonial times, but by about 1800, the newly independent nation had become settled and comfortable enough to take an interest in formal gardens in the European tradition. Many of the most common urban landscaping species—Lombardy poplar, Norway maple, and plane tree, among many others—were introduced to North America around the beginning of the nineteenth century by enthusiastic gardeners. These early introductions were mostly European in origin and reflected the cultural history and economic ties of the young republic. Among these introductions were species destined to become some of the most widespread invasive pest plants in North America.

It wasn't until around 1850 that the United States began to expand its political (and botanical) focus to the Far East. Before the mid-nineteenth century, only a handful of plants native to China and Japan (probably most prominently the gingko tree) had seen Western shores. But once Asia was forced to open itself to the outside world—China following the Opium War of the early 1840s, and Japan with the arrival of Commodore Matthew Perry in 1853—the trickle of oriental ornamentals soon increased to a flood as botanizers followed businessmen and missionaries into the vast uncharted territories of eastern Asia. Today, many of the most common ornamental species in cultivation in North America—forsythia, for example—are of Asian origin. So too are a large portion of the worst plant invaders on the continent, from tree of heaven ("tree from hell," a naturalist friend of mine calls it) to multiflora rose.

Even in the early days of the American republic, ornamental gardens were not solely the purview of wealthy city dwellers. Ornamental plants traveled with pioneers as well, and the doorsteps of settlers bloomed with flowers and shrubs from around the world long before all the forests and prairies were tamed. Not only did people carry seeds and live plants with them as they moved westward—they also ordered plants from an ever-expanding array of mail-order nurseries that offered everything from wisteria seeds to tulip bulbs. Gardens were often considered fashion statements, and like most fashions, they were continually driven to the newest and the most unusual. In this respect, gardens in America played a similar cultural role to their counterparts in Europe.

But more fundamentally, nineteenth-century Americans—urban and rural dwellers alike—saw ornamental plants as a powerful symbol of civilization's triumph over wilderness, and as a consequence, garden plants carried weight above and beyond their value as adornments. As one historian has noted, carefully maintained gardens were "quiet yet very public reminders of middle-class values and respectability." They served to differentiate between the well-kept and the unruly, the civilized and the wild—and by implication, between the introduced and the native. Though some indigenous plants (particularly trees) were used in landscaping and gardening, American culture showed a clear preference for exotic ornamentals from the start, and the introduction of new plants has long been almost universally regarded as a broad public good.

In a sense, ornamental plants are still more cultural symbols than they are living organisms in the mind of the American public. Most people consider them an extension of their homes, and it always seems to catch people by surprise when one or another species turns its back on civilization and goes wild. Even though about half of all invasive pest plants in North America are escaped ornamentals, it is still difficult for many people to learn to see a plant that was selected for its beauty as a harmful invader. Somehow it feels like a betrayal whenever the garden of dreams turns into a nightmare.

## The Woods Invaded

There is nothing better than spending a hot August afternoon ankle-deep in a marsh. And when that marsh is in the middle of a patch of woods—well, even better. So when my friend Jill Baum said she needed help with her thesis research in Picnic Point Marsh on the University of Wisconsin–Madison campus, I didn't have to spend very long thinking it over.

Jill was waiting at the entrance to Picnic Point with meter tape and clipboards in hand. As we headed down the path toward the marsh, I asked her to explain what her research was about. "Picnic Point Marsh has been shrinking for decades," she answered, "and nobody really knows exactly why." Trees and shrubs have been taking over what was once open wetland and squeezing out the native marsh vegetation in the process. Jill's plan was to document current vegetation patterns in and around the marsh as a baseline for future research on the mechanisms and causes of change.

But I wasn't so much interested in the marsh as I was in the woods that surrounded it. As we began to move from marsh into woods along the first sampling transect, I started paying close attention to what Jill found at ground level. As soon as we left the marsh, the change was dramatic: In the space of a few yards, the ground layer essentially disappeared. There was no ecotone, no transition between open marsh and closed woods. Meter 57. "Nothing." Meter 60. "Buckthorn seedling." Meter 66. "Nothing." Meter 74. "Nothing." The edge of the woods was a wall of tangled shrubs. And underneath them, the ground was almost completely bare.

It seems that this patch of woods has in recent history changed just as dramatically as the marsh. According to John Curtis's *Vegetation of Wisconsin*—a 1959 publication that is the standard reference on the subject—woods like these should have anywhere from ten to forty native species of understory herbs and grasses and a scattering of tree seedlings. We found little but bare ground. Unlike the marsh—where the causes of decline are so mysterious as to warrant systematic research—the reasons for the degradation of these woods are in broad terms quite clear. Heavy human use, the absence of fire, and the small size of this isolated patch of woods have all played a role in the erosion of understory diversity. The single most important element, though, is the overwhelming dominance of honeysuckle and buckthorn, the two invasive nonnative shrubs that spring up at the very edge of the marsh and blanket the understory beyond.

Honeysuckle and buckthorn were among the earlier ornamental introductions to the North American continent, and like most imported plants, they were greeted with nothing but praise. What we commonly call honeysuckle is actually a complex of several closely related species introduced from Asia and Europe over the course of the nineteenth century that tend to hybridize with each other. Buckthorn is the name generally applied to two distinct (but nonhybridizing) species, both native to Europe. (There are native North American buckthorns and honeysuckles as well, but they are unfortunately much less common than their introduced cousins.) Both honeysuckle and buckthorn were (and still are) used widely as landscape plants, and honeysuckle has been planted extensively to enhance habitat for various bird species as well. (Their dense foliage is often inhabited by low-nesting bird species, and their berries provide a reliable food source for fruit-feeding birds.) The apparent benefits of these shrubs to wildlife only increased their popularity, and with the help of fish and game agencies, nurseries, and private individuals, honeysuckle and buckthorn spread across the country at a tremendous rate.

The Campus Natural Areas are not the only conservation lands in Madison that have been invaded by buckthorn and honeysuckle. These species are also overwhelmingly dominant in many sections of the University of Wisconsin Arboretum's 1,200 acres. A study by University of Wisconsin graduate student Liling Li showed that honeysuckle and buckthorn are in fact the most numerous and most widespread woody species in the wooded natural areas in and around Madison. While their presence is by no means the only cause of loss of understory plant diversity in wooded areas in the region, they are certainly major contributors to its decline.

The principal competitive advantage that honeysuckle and buckthorn have over native species seems to be their long season of active growth. Both honeysuckle and buckthorn leaf out earlier in the spring and keep their leaves until later in the fall than any native deciduous shrub or tree. (Buckthorn commonly has green leaves until after Thanksgiving in Madison, and I have seen honeysuckle bushes hold their leaves until early December.) Long-lived foliage confers several benefits to these species. First, it allows for a longer photosynthetic season, which may help honeysuckle and buckthorn

Buckthorn and honeysuckle often form dense stands in the midstory of native forests, preventing the regeneration of native tree species. (Jason Van Driesche)

grow faster than competing native shrubs and young trees. Research has shown that nearly one-third of the annual carbon gain of both honeysuckle and buckthorn occurs before one of their common native competitors (gray dogwood) even leafs out. Second, once honeysuckle and buckthorn are firmly established on a site, they tend to shade out almost all seedlings of other species. Spring ephemeral wildflowers are particularly vulnerable, because honeysuckle and buckthorn effectively close the window of sunlight that normally exists between the beginning of the growing season and the leafing-out of the first native trees and shrubs. Essentially, these invaders take over and maintain control by working overtime—something few native species are evolved to do.

Buckthorn and honeysuckle have somewhat different habitat preferences, but they are generally found together in any given community. Their impacts are in a broad sense essentially equivalent as well, and most land managers treat honeysuckle and buckthorn as a single problem. (Some informally merge the species, referring to them collectively as "bucksuckle.") They tend to form dense stands up to two stories high, and under some circumstances can almost entirely eliminate native shrubs and saplings. A recent study suggests that extensive thickets of these species may even act as "ecological traps" for at least two bird species. They found that robins and wood thrushes nesting in honeysuckle and buckthorn experienced significantly higher predation than those nesting in comparable native shrubs and trees—suggesting that invasion by these shrubs may represent a net loss of habitat quality for some birds. Native plants and animals almost all suffer wherever these two invaders take hold.

But not all habitats are equally vulnerable to invasion by honeysuckle and buckthorn. There is general agreement among ecologists who have studied these species in the Madison area that these invaders are a problem largely in urban natural areas. Though the seeds of both species are bird-dispersed (and are therefore carried far and wide), there are large forested areas where buckthorn and honeysuckle have yet to establish themselves. For instance, I've hiked for miles in Devil's Lake State Park, a 10,000-acre forested area 40 miles northwest of Madison, without seeing anything more than an occasional lone buckthorn or honeysuckle plant. The biological attributes of the invaders are the same—but clearly the ecological attributes of the habitat are not.

There are a series of characteristics common to most urban natural areas that distinguish them from rural natural areas and make them especially vulnerable to large-scale invasion. Some of these characteristics are a product of the history of the natural area itself; others are related to the status of the lands that surround it. All of these characteristics place significant constraints on the ecological future of places such as the Campus Natural Areas or the Arboretum.

First, all urban natural areas are (of course) in cities. High concentrations of people mean high concentrations of ornamental plants. In general, there are both a wider variety of nonnative species and more individuals of any given species in urban than in rural areas. This translates into a higher probability that one or another nonnative species will find its way into an urban natural area.

Second, urban natural areas are generally quite small. Size matters because as the size of a natural area decreases, a larger and larger proportion of its habitat will be within a short distance of the edge of the natural area. Since urban natural areas are often bordered by landscaped yards and gardens—common sources of invaders—this kind of edge habitat is almost always weedier than interior habitat. Therefore, the smaller the natural area, the more likely it is to suffer from such "edge effects."

Third, most urban natural areas are recovering from heavy past disturbance. The Arboretum's prairies and woods are not untouched remnants of intact native ecosystems; they are farm fields replanted with prairie species and woodlands still recovering from a half century of grazing and plowing. While the general appearance of forest and prairie are relatively easy to recreate, restoring proper function to a system that has been largely destroyed is a much more delicate and difficult business. Invaders are more likely to find a foothold in a system whose basic functioning has changed so drastically in such a short time.

Fourth, portions of many urban natural areas were at some point in the past actively planted with nonnative species. The Civilian Conservation Corps planted honeysuckle in the Arboretum in the 1930s to "improve" bird habitat. Portions of the Campus Natural Areas are former estates whose extensive landscaping is still visible amidst the native vegetation that has since recolonized the site. Establishment is often the largest hurdle for an invader, and any nonnative species that is planted and protected by humans is more likely to become invasive than is one that did not enjoy such an advantage.

Fifth, urban natural areas are sometimes used as detention basins for municipal stormwater, a practice that can bring in many troublesome invasive species. For instance, the Arboretum's Greene Prairie—a beautiful, highly diverse wet prairie restoration—is being obliterated by nonnative reed canary grass. The invasion started in the outwash from a drainage pipe that carries sediment and seeds in from a degraded marsh on the other side of the highway and has been spreading with the stormwater ever since.

Finally, urban natural areas are often heavily used for recreation. Even the most careful of hikers are likely to bring in seeds of nonnative species on their clothing and shoes or on the tires of their cars. For this reason, invasive plants—particularly new invaders—are often concentrated along trails. In addition, the soil compaction and erosion caused by hikers and mountain bikers can stress an ecosystem, making even a relatively intact area more vulnerable to invasion.

But this is not all. The same characteristics that make urban natural areas both more vulnerable to invasion and more likely to be invaded than rural wildlands also tend to make controlling invaders in urban natural areas rather controversial. Heavy public use means that urban natural areas are under intense public scrutiny, and when management plans call for removing species that people grow in their own gardens, public reaction is often less than favorable. Though land-use history and environmental context are largely what drive invasions in urban natural areas, a cultural reluctance to see invasive plants (or any plants, for that matter) in anything but a positive light is probably the biggest hurdle to controlling the impacts of such invasions in places like the Arboretum and the Campus Natural Areas.

## To Cut or Not to Cut?

It was a good thing they had booked a large hall for the second Campus Natural Areas public forum. There were nearly a hundred people in attendance, and many of them had quite a lot to say. Following several controversial management decisions made about a year and half earlier, plans for ecological restoration in the Campus Natural Areas were on hold and their future under review. These people all wanted a voice in what happened next.

"I've heard a rumor that there are plans for cutting down the woods on Picnic Point," an audience member said. "Is this true?" Many others at the meeting were probably wondering the same thing. In the summer of 1998, the Arboretum had initiated a restoration project in Frautschi Point Woods, another unit of the Campus Natural Areas. Most local residents had no idea that any management action was planned until a little less than an acre in the middle of the woods was summarily stripped of trees and brush and seeded with prairie species. What had been continuous wooded cover was now interrupted by something many people were calling a clear-cut, and many local residents were quite upset. One wrote a letter asking, "Why the cutting? Why the destruction? Aren't there enough nondestructive conservation projects to keep people busy?" The university received a number of letters and calls protesting the project, and both restoration work and research in the Campus Natural Areas abruptly came to a halt.

Many of the people who opposed the cutting at Frautschi Point were afraid that this same kind of management would soon be applied more broadly across the Campus Natural Areas and objected to the kind of future that such management would imply. Some were concerned about the impacts of the removal of shrubs—largely honeysuckle and buckthorn—on native bird populations, arguing that thrushes, warblers, and other midstory-dwelling birds would be deprived of habitat and their numbers would drop if woods were converted to a more open vegetation type. Others expanded on this point to argue that the restoration paid almost no attention to animals and instead focused almost exclusively on re-creation of native plant communities. Tom Powell, a county supervisor, was also concerned that erosion and siltation into the lake would increase as a result of the cutting of honeysuckle, buckthorn, and other shrubs. Many others simply protested the cutting of trees and the loss of wooded cover in general. Glenda Denniston, a longtime Campus Natural Areas volunteer, echoed the views of most of the project's opponents when she argued for the necessity of preserving wooded areas in a rapidly growing metropolitan area. "I want places that are quiet, that are private, that you can go into and be alone," she said. "We need places where people can find peace." To her mind, the cutting on Frautschi Point clearly didn't qualify.

Though the loss of tree cover at Frautschi Point was what initially drew criticism in the Campus Natural Areas, protest soon spilled over from this specific management action to the management plan as a whole. This was an unfortunate turn of events, because the plan itself was in many respects quite sound. It was written by former Arboretum ecologist Virginia Kline and Arboretum staff member Brian Bader to reflect and build on the strengths of the existing landscape, restoring native woodland wherever it was found and encouraging a mix of prairie, savanna, and woodland in places where open land or invasive species dominated. The long-term goal of the plan was to re-create a "diverse landscape of prairie, savanna, woods and wetland reminiscent of the presettlement landscape." According to Virginia Kline, the plan did not call for the elimination of any mature trees at all. The only trees to be removed were young "weedy" invaders (a mix of native and nonnative, fast-growing, short-lived trees) that had sprung up over the last few decades in areas that had previously been open land. Nonnative brush—largely honeysuckle and buckthorn—would be removed only as fast as it could be replaced with native midstory species. Controlling erosion was a major focus of the plan as well, and Kline was confident that if it were done carefully, restoration of a robust native mid- and understory would reduce soil loss in places where the ground is now bare underneath dense thickets of honeysuckle and buckthorn.

In the broadest sense, the ecological intent of both the plan as a whole and the cut at Frautschi Point in particular was to approximate as closely as possible an ecosystem unaltered by the many stresses that Europeans have over the last few centuries imposed on the land. Using presettlement conditions as a benchmark for designing management prescriptions and measuring success is quite common among restoration projects, and in many ways this approach makes good ecological sense. The time of settlement may or may not represent the high point of biodiversity and ecosystem integrity in North America, but it is the best target that historical records provide, and re-creating the conditions of that point in time is a good first step in bringing back what has been lost in the years since.

The problem was, the work at Frautschi Point itself was undertaken with *nothing but* these ecological goals in mind—even though the plan as a whole emphasized broader social goals as well. The decision to cut was made without taking into consideration the needs and desires of the people who cared about and used that piece of land. This approach ignored the reality that in an urban natural area such as this one, the *context* of the area to be restored has changed so dramatically that presettlement conditions are by themselves no longer adequate as the sole standard for restoration.

At the time of European settlement, the savanna and woodland and prairie that is now the Campus Natural Areas was surrounded simply by more savanna and woodland and prairie. Now it is surrounded by a city of 200,000—a context that places a fundamentally different set of constraints and pressures on the Campus Natural Areas (both ecological and social) than those that existed at the time of settlement. This change of context is precisely why large-scale savanna and prairie restoration is so desperately needed—and why the cutting drew such strong protest. The project essentially ignored the realities and the demands of the present, constructing instead a definition of "natural" that looked only to the ecology of the past. But restoration work needs to take more than soil type and surveyors' records into account—particularly in an urban context where so much has changed.

The demands of the present are relatively simple: People don't want the Campus Natural Areas to change too much. Or at least not too quickly. And they want to be involved in shaping the process of change. Leo Garofalo, a graduate student who lives near Frautschi Point and often goes for walks there, simply wanted to see less human intervention in the ecosystem. "We need to limit the amount of change we cause," he said. "The important thing is to maintain the quality of that wooded space." By "quality" he meant essentially that he didn't want to go for a walk in the woods and feel like he was in the middle of a logging camp or a construction site. Glenda Denniston was equally concerned about the nature of the changes under way, but given her long-standing involvement in alien species control and restoration of native woodlands, she was more concerned with the abruptness of the changes caused by the restoration work than with the amount of change per se. What these and many other people who protested the cut had in common, though, was a deep sense of uneasiness about the future of the Campus Natural Areas.

The basic argument of many of the people who protested the cut at Frautschi Point rested on one assumption: that there is little direct conservation value in creating a series of postage-stamp-sized natural communities on degraded land in the middle of a city. And in a sense, they are right—like most urban natural areas, the Campus Natural Areas are too small, too isolated, and too heavily used to have much ecological value. Moreover, whether the university actually does any of the restoration called for in the management plan has relatively little immediate bearing on the overall conservation status of prairie and savanna species and communities across the region as a whole.

But while the direct ecological value of whatever happens to this strip of woods is relatively insignificant, the *social impact* of the Campus Natural Areas' ecological status is tremendous. If the Campus Natural Areas are managed for maximum wooded cover, then most of the people who visit the natural areas will end up knowing only one kind of wild place: the kind that is there now. And given the tremendous stresses that this system is already under, this kind of management is likely to be too passive to keep these woods from sliding even farther toward a lowest common denominator of a handful of aggressive tree species and a midstory of honeysuckle and buckthorn—a kind of "natural" that is good for hiding the city skyline and not much else.

Such management creates a kind of ecological myopia that inevitably leads to a progressively narrower vision of the past. If maximizing tree cover becomes the standard management practice for the Campus Natural Areas, more and more urban residents will live their lives without any real, concrete knowledge of the full spectrum of communities that are native to the region as a whole. Such an erasure of ecological history is a high price to pay, for urban residents will only care enough to support restoration of native communities in the places where it really matters ecologically if they are given the opportunity to see and care for intact, invader-free prairie and savanna and wetlands (and woods!) in their own backyards.

Public reaction to the work done at Frautschi Point has made it clear that restoration of native ecosystems—and in particular, control of invasive species—is as much a social challenge as an ecological one. While the restoration work was doing a good job of lessening the impacts of buckthorn, honeysuckle, and other major invasive species, it clearly is not a useful model for restoration in the Campus Natural Areas as a whole. In fact, because of the controversy over this first cut, all restoration work in the Campus Natural Areas has been suspended—including the control of buckthorn and honeysuckle. An approach to restoration and invasive species control that leaves out people clearly does not work.

Fundamentally at issue in this debate is how to perpetuate that essential quality that we call "nature" in the middle of a city. Restoration is predicated on the idea that calling a place "natural" entails providing a counterbalance to the human stresses (past and present) that have been placed on the land—that a degraded site can and should be actively managed. More passive approaches to management assume that "natural" means working within the context of whatever nature is present on a site. But this is a debate that will never be resolved; for not only are these two views antithetical to each other, they are also arguing over something that in an urban context no longer exists.

These scraps of woods and fields will never be "natural," no matter how carefully we compensate for—or how thoroughly we attempt to remove— the weight of human presence. But they can in many ways be made useful. If we manage them well, they could eventually become a pleasing and welcoming environment for plants and animals and people of many kinds. If they are managed poorly, they will become a weed patch. In any case, they will always be a product of human design, whether we admit it or not.

The only way out of this dead-end debate is to let go of arguing about what constitutes "natural" and to design a future that will work for everyone. "The idea now is to go slowly, with lots of education along the way," explained Bob Ray, the chair of the task force that has been charged with reevaluating the future of the Campus Natural Areas. Through planning sessions and public meetings and many hours of deliberation, the task force has begun a process of setting measurable, specific goals in a language everyone can agree on. "You can't just say, 'We're going to restore it,'" Ray said. "People will look at you and ask, 'Why?'" This is the question that now must be answered.

# The Gift of Meaning.

When the first trees came down in the Arboretum's Wingra oak savanna in late winter 1992, many of the people who lived across the street from the Arboretum in the Dudgeon-Monroe neighborhood were outraged. Wasn't the Arboretum supposed to protect trees? Instead, it was cutting down cottonwoods over 2 feet in diameter. In short order, angry protesters gathered nearly 200 signatures on a petition calling for an end to the cutting.

When they presented the petition to their city alderperson, they were advised to take the issue up with the neighborhood association. The group soon discovered, though, that the association and the Arboretum were in communication about the issue—and that the association had already endorsed the cutting as part of a larger restoration project that would over the course of several years remake the character of the entire site. A few months earlier, association co-president Henry Hart had read in the paper that the Arboretum was planning a restoration project for the Wingra oak savanna and decided to talk with Arboretum director Greg Armstrong about how the neighborhood might become involved. Armstrong had already been thinking that incorporating the opinions and concerns of local residents was going to be necessary to the success of the project, and the two arranged for an Arboretum staff member to speak at the next meeting of the neighborhood association.

About forty local residents attended the January 1992 neighborhood meeting, and Molly Murray of the Arboretum explained the proposed project. When the first European settlers began arriving in the 1840s, she said, they found an open savanna of huge bur oaks and a great diversity of wildflowers and grasses across most of what is now the Dudgeon-Monroe neighborhood. One hundred and sixty years later, some of the old oaks still survived, but lack of fire had allowed fast-growing trees and invasive shrubs to take over the understory. The Arboretum's proposal was to clear out the invaders—mostly buckthorn and honeysuckle—and reintroduce fire in order to restore the community as a whole and bring back the understory species that had been lost. A spirited discussion ensued, but there really was no opposition. The association voted yes.

So when the cutting began in early 1992, the outburst of protest and the petition that was circulated surprised neighborhood leaders. After all, the decision to join in the project had been made openly and democratically. Even so, the association's leaders were willing to reopen the discussion to allow for the participation of the protesters and other interested local residents. They set up a public meeting on the site of the cutting and invited

Oak savannas are characterized by a highly diverse understory of grasses and wildflowers. (Richard Henderson)

anyone concerned to come and discuss the project. Molly Murray and Brock Woods, the Arboretum staff members in charge of the restoration, explained how cottonwoods and other aggressive, fast-growing trees were killing the lower limbs and stunting the growth of the site's large bur oaks. According to Molly Murray, "The turning point came when we compared the growth rings of a cottonwood with those of a relatively small oak tree that had been toppled in a windstorm. The huge cottonwood was only 35 years old, but the oak was more than 135 years old." Like the people who attended the original meeting, the protesters saw the need for some careful cutting and dropped their petition to stop the project.

This cycle repeated itself two more times over the next few months. As more neighborhood residents became aware of the cutting, a new round of protest would surface, and the neighborhood association would arrange another on-site meeting with Arboretum staff. Gradually, the community as a whole came to understand both the nature and the consequences of the changes the savanna had undergone in the century and a half since settlement. Ecological history took its place in the debate, and the scope of what the neighborhood saw as possible on the site began to expand.

But what happened at Wingra was more than just one-way Arboretum-to-community education. Through the relationship that had been estab-

lished between the neighborhood association and the Arboretum, local residents participated in both the planning stages and the on-the-ground work
for the project. According to Henry Hart of the neighborhood association,
the Arboretum modified its plans several times in response to concerns
raised by local residents. For example, when several people expressed reservations about the loss of privacy and solitude along Arboretum paths once
honeysuckle and buckthorn were removed, Arboretum staff agreed to take
out the invasive shrub cover in stages and replace it with strategically placed
clumps of native shrubs. (Loss of honeysuckle and buckthorn thickets was in
fact the primary concern of one of the most vocal of the project's early opponents—a local resident who, after several meetings with Arboretum staff,
became one of the restoration effort's most active volunteer leaders.) This
kind of give-and-take afforded local people a sense that the project took
them into account—that restoration was as much about people as it was
about grass and trees.

Another kind of local ownership of the project grew out of a series of
twice-monthly volunteer work parties that the neighborhood association
and the Arboretum began to conduct in the summer of 1992. Volunteers cut
brush, planted prairie grasses and forbs, and collected native seed for future
plantings. Every work party also featured an installment in what came to be
known as the "tailgate lecture series," in which Brian Bader, the Arboretum
staff person who took over leadership of the project in 1995, would lead a
discussion on some broader idea related to the day's work. The brush cutting
and seed planting eventually coalesced into what the Arboretum now calls
its Earth Partnership program. It has since attracted hundreds of volunteers
with its mission of restoring not only the landscape but also the relationship
between the land and the people who inhabit it.

The high point of each year's work at the Wingra oak savanna is the
spring Restoration Celebration, where the neighborhood gathers at a stone
fire ring that was built years ago in the middle of the savanna for a potluck
dinner, a bonfire, a guest speaker, and a discussion of what's in store for the
year to come. As of early 1999, nearly half of the site had been cleared of
invasive species and planted to native grasses and wildflowers, and the
savanna was beginning to reappear. But perhaps even more important than
the resurgence of the native community itself was the emergence of a strong
restoration ethic in the neighborhood and a deep understanding of the
enterprise of restoration. As Brian Bader and Dave Egan of the Arboretum
described it, "Those who regularly attend the volunteer work parties now
understand the restoration process—what works and what doesn't work.
They know what the Arboretum is trying to accomplish and they are able to

communicate it to passersby. Some are even able to spot problems or suc-cesses before Arboretum staff do." This was restoration in the fullest sense of the word.

Restoration of this kind has its roots in the Arboretum itself. Some of the earliest restoration work in the country began right here under Aldo Leopold, a professor at the University of Wisconsin in the 1930s and one of the grandfathers of modern conservation. But according to Bill Jordan, founding editor of the Arboretum-sponsored journal *Ecological Restoration* (formerly *Restoration and Management Notes*), it is only since the 1980s that restoration has come to be defined as a social as well as an ecological enterprise. In fact, Jordan said, restoration was for many years almost a dirty word in ecological circles. The term was associated with the half-hearted reclamation efforts that typically followed strip mining, and calling a place "restored" was tantamount to calling it fake. Under Jordan's leadership, the Arboretum began to redefine restoration as both a rigorous ecological discipline and a valuable educational tool. During the late 1980s, Jordan began experimenting with restoration-oriented tours in the Arboretum, and in 1991 he and Arboretum ranger Brock Woods teamed up to launch a restoration-based education program, ultimately merging what had been three separate programs—education, volunteers, and restoration—into one. As a result, the development of the approach that made the Wingra oak savanna project so successful was well under way at the Arboretum several years before the first trees and brush were cut along Monroe Street.

Dave Egan, another of the editors of *Ecological Restoration*, explained that the Arboretum's approach is so effective because it is just as much about peo-ple as about native ecosystems. "First of all, you have to throw out all notions of 'natural,'" he said. "Put in its place 'historic'—in the sense that both the land and the people have a history." At its best, he explained, restoration connects human history with ecological history by helping people expand their knowledge of the landscape to include something more than what they have personally seen and experienced. Only the oldest residents of the Dud-geon-Monroe neighborhood might still have remembered when the Wingra oak savanna was actually a savanna. The rest of the neighborhood had never seen anything but a tangle of cottonwoods and oaks and honeysuckle and buckthorn. The most basic task of the restoration project was not to restore the savanna itself, but to restore the *possibility* of a savanna in the imagina-tions of the people who knew it only as woods.

This kind of expansion of imagination draws very different lines on the landscape than does the debate over what is and is not "natural." Rather than arguing over whether Native American fire management made savan-

A restored savanna just outside of Madison, Wisconsin. (Jason Van Driesche)

nas unnatural, or whether species introduced by Europeans should be treated differently than those that were here before settlement—in short, rather than arguing over what was human and what was not—the Wingra restoration project based its judgments and its distinctions on one deceptively simple question: What will give this place the most meaning?

David Mollenhoff, a local resident and author of a history of the Madison area, addressed this question in a speech at the first Restoration Celebration in May 1992. He recounted the history of the Wingra site, from the retreat of the glaciers to the arrival of Native Americans to the settlement of the area by Europeans. He wove in with its human history an account of the changes in its vegetation, from boreal forest to hardwood forest to oak savanna. He showed how the two histories had for millennia been connected—and how the two were still connected through the changes that modern humans have caused, from fire suppression to the introduction of new species. He highlighted the discontinuity between recent history and deep history, both human and ecological, and allowed the question to arise in the minds of his audience: How can meaning be restored to a place cut off from its past? He left with a word of thanks for those who had already begun to find an answer. "I know I speak for many when I express my gratitude to the small group of dedicated people who have worked long and hard to make this project a reality," he said. "What you have done is to give us a gift of great value. You have given us the gift of the past. You have given us the gift of place. You have given us the gift of meaning."

## Extending the Boundaries

What the Arboretum did with the Wingra restoration project was effectively to extend its functional perimeter beyond the line on the map that constitutes its official administrative boundary. National Park Service researchers Christine Schonewald-Cox and Jonathan Bayliss described this functional perimeter in conceptual terms as a "generated edge" that shrinks and expands in response to the relative balance of forces (both positive and negative) that operate on natural areas. If negative cross-boundary influences—small size, heavy use, history of disturbance, or even public opposition to management actions—are greater than whatever positive influences management can bring to bear, the generated edge of the natural area will fall somewhere inside its official boundary, and the problems related to nonnative species invasions are likely to increase, particularly along the perimeter of the natural area. But if positive cross-boundary influences can be augmented to the point where they compensate for negative influences, the generated edge will expand beyond the boundaries of the natural area, and the invasive species problems will decrease.

Of all the forces that affect the integrity of places like the Arboretum, public support has perhaps the greatest influence on their condition through its effect on such natural areas' generated edge. With this idea in mind, Dave Egan and Steve Glass (the Arboretum land-care manager) expanded on Schonewald-Cox and Bayliss' framework by proposing a program of "watershed volunteers" as a means of developing a working relationship between an urban natural area and its neighbors. Their working hypothesis was that giving local residents a stake in restoration efforts creates a social buffer that is in many ways even more effective than the physical buffer of more land. A quarter-mile-wide swath of undeveloped land along the outer perimeter of a natural area might provide a measure of protection from invasive species, but a cadre of fifty dedicated, long-term volunteers is even better. As threats to its natural communities increase, the Arboretum will have to rely more and more on "watershed volunteers" and other social buffers to counter the negative consequences of the fact that it is in the middle of a city.

In fact, the Arboretum has already begun to do just that. Glass and Egan describe how the Arboretum has augmented its control effort for garlic mustard (a highly invasive understory weed) with a program that equips neighborhood volunteers with educational materials related to garlic mustard and sends them door-to-door in their neighborhoods to raise awareness across the city about this pest. The Arboretum also did a series of press releases and public events designed to alert people throughout the Madison area to the threat of garlic mustard. The results were remarkable. As Egan and Glass describe it, "Garlic mustard became a general topic of conversation. . . . We got literally dozens of calls from citizens throughout the city who discovered that they, too, had a garlic mustard infestation and wanted to know how they could control it." Though garlic mustard is far from being eradicated at the Arboretum, finding a solution is significantly more likely now that the Arboretum is not the only one addressing it.

This, then, should be the ultimate goal of urban restoration: to create a positive, mutually beneficial relationship between urban residents and urban natural areas that extends the functional boundaries of natural areas beyond their official perimeters to include the city and even the region as a whole. If through restoration work enough people come to see the whole landscape as an entity with a history to be respected and a future to be created, then dealing with invasive species will simply become part of the daily lives of the inhabitants of every urban landscape. Cities will change from a source of invasive species problems to a source of solutions. And the garden of dreams will give way to a garden of meaning.

CHAPTER 13

# Bringing It Home
## *Public Awareness of Alien Species Impacts in Hawai'i*

If you didn't know exactly where it was, you'd never find The Nature Conservancy's main Hawai'i office. "You said what . . . —*Smith* Street?" the cab driver asked. I nodded and told him I thought it was in the old part of Honolulu. Twelve dollars later, we pulled up in front of a door tucked between two storefronts halfway down a side street. The office was marked by a small wooden sign, about a foot and a half square, that read "The Nature Conservancy of Hawai'i."

This unassuming facade betrays little of what goes on inside the Conservancy's office. The reception area feels more like a campaign headquarters than a place to sit and read a magazine. There are posters for the Conservancy's series of conservation programs on the walls, newsletters lined up on the coffee table, and brochures for current campaigns in a rack on the wall. The Conservancy's office is one of the focal points of an ambitious statewide campaign to change the way people think about nonnative species. Though the public may not be aware of it, public awareness begins in places like this.

Grady Timmons is the communications director for the Hawai'i chapter and spends much of his time promoting public awareness of the impacts of alien invasive species. He invited me into his office, and we sat down amid stacks of press clippings and planning documents and mock-ups of campaign posters. I asked him to describe the state of public awareness about alien pest species in Hawai'i. Grady quickly made it clear that there was no single answer to my question—there exists more than one "public" to deal with, he explained, and a different level of awareness that is necessary within each. Defining these many audiences and designing communications strategies to reach each of them has become a major thrust of the Conservancy's work in Hawai'i.

271

MAUI

KAHULUI AIRPORT

Miconia infestations

HANA

Kipahulu Valley

Haleakala National Park

AREAS WITH MICONIA INFESTATIONS

PROTECTED AREAS

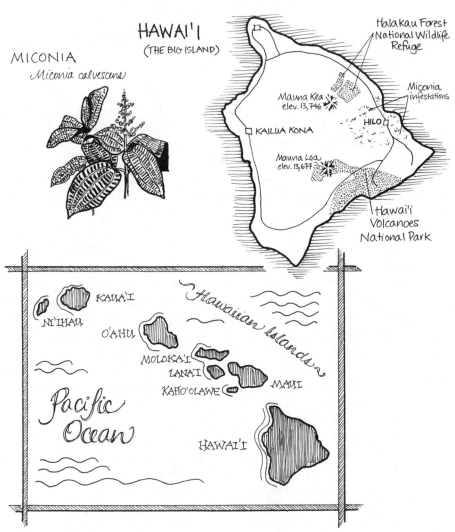

MICONIA

*Miconia calvescens*

HAWAI'I
(THE BIG ISLAND)

Halakau Forest National Wildlife Refuge

Miconia infestations

Mauna Kea
elev. 13,746

KAILUA KONA

HILO

Mauna Loa
elev. 13,677

Hawai'i Volcanoes National Park

Hawaiian Islands

KAUA'I

NI'IHAU

O'AHU

MOLOKA'I

LANA'I

KAHO'OLAWE

MAUI

Pacific Ocean

HAWAI'I

Awareness in Hawai'i about nonnative species invasions already seemed to be high. Grady pulled out a 1996 Conservancy-sponsored survey that indicated that once the concept of "alien pest species" was explained to them, nearly 60 percent of Hawai'i residents surveyed agreed that such invaders were "a very serious problem facing Hawai'i." But how many respondents actually understood the problem and their role in it? Most people express only a vague sympathy for the issue, largely because it is something that is easy to agree with. (After all, who's going to argue for the introduction of anything described as a "harmful pest"?) The real test of public awareness is whether it leads consistently to public action—and on this count, Grady said, while Hawai'i has made a start, much more work is needed.

On the flight from the mainland earlier that same day, I had spent a half hour or so talking with a flight attendant about the islands. "Hawai'i is a unique place," she told me. "There's a sweetness about life there." Like many visitors and residents, she clearly saw value in the fact that Hawai'i is a world apart and wanted it to stay that way. Her fondness for the islands predisposed her to care about the threat of alien invasive species—but since she knew little about the impacts of current pests or the ways in which new invaders are introduced, her concern about the issue was entirely passive. This kind of support registers in opinion polls, but is not likely to generate letters to legislators. What she needed was a reason—or lots of reasons—to make her concern active.

This flight attendant is exactly the kind of person whom public awareness efforts are trying to reach. And so is the cab driver who brought me to the office, and the elderly lady who runs the hostel where I'm staying, and the young guy who sat next to me at the lunch counter down the street. They probably all think Hawai'i is a pretty nice place, but don't understand the degree to which the things they like about Hawai'i are threatened by nonnative species invasions. Until they do, invasive species will remain a back-burner issue, and Hawaii's native ecosystems will continue to fall apart.

Increasing public awareness about the consequences of nonnative species invasions was looking much more difficult than I had imagined. Not only does it require an understanding of these different "publics" and their often-divergent views on what qualify as important issues; it also demands that these perspectives come together in a common sense of concern about the threat of nonnative species in general. Trying to galvanize these audiences individually won't work. If the conservationists worry only about new weeds invading forests and the farmers simply look out for new pests that might attack their crops and the hotel owners are concerned only about invasions of biting sand flies on their beaches, none of them will be able to adequately protect their interests from new invaders. Working together, however, they may have the power to persuade the entire state—individuals, businesses, and government—to take the threat of nonnative species invasions seriously.

So this is the point of all the buzz and activity in this out-of-the-way corner of Honolulu. The Conservancy and its many partner organizations are attempting to galvanize a unified front of so many interests that concern about nonnative species impacts cannot help but rise statewide. Their "Silent Invasion" campaign is young, and the changes they hope to instigate are long-term in nature. But if it is successful, this approach to preventing and fighting invasions may finally make nonnative species battles winnable.

## The Silent Invasion Campaign

"I always wanted to do something that mattered," Gary Sprinkle said as he put the videotape into the player. Raising awareness about alien pest species is where he got his chance. As news anchor for KITV, one of Hawaii's biggest television stations, Gary is a well-known public figure in a highly visible forum. He began working for KITV in the early 1990s and used that forum from the start to draw attention to the beauty and the vulnerability of Hawaii's native flora and fauna. Since 1995, his focus has expanded to include the threat of nonnative species invasions. In a series of short spots and longer specials, Gary and KITV have played a significant role in publicizing the impacts of nonnative invaders.

The narrator's voice was hard-edged, and a snake slithered across the screen as he spoke. "If you're not scared of brown tree snakes becoming established in Hawai'i," he warned, "then you don't know what they have done to Guam." Over the next fifty seconds, the tape made three essential points: Brown tree snakes have emptied Guam's forests of birds, they bite people who are asleep in their beds, and they cause more than a hundred hugely expensive power outages every year. That the same thing could happen in Hawai'i is driven home by the closing image of a plane on Guam taxiing for takeoff for Hawai'i—a plane that could have a snake stowed away on it.

KITV has produced more than a dozen public service announcements like this one. There are several versions of each spot, and the series airs during prime-time programming. These spots have had a significant impact on public awareness of invasive species threats. "When you mention the brown tree snake now," Grady Timmons said, "a very high percentage of people know what you are talking about."

But the Silent Invasion campaign is larger than just the KITV spots. The statewide effort against invasive species began with a 1992 report by The Nature Conservancy of Hawai'i and the Natural Resources Defense Council entitled "The Alien Pest Species Invasion in Hawaii: Background Study and Recommendations for Interagency Planning." The report described the mechanisms that were then in place for preventing new invasions and for controlling existing pests and highlighted gaps in the system as a whole. It then offered a list of "next steps" for making prevention and control efforts more effective statewide.

First was the organization of a multiagency system for preventing, intercepting, eradicating, and controlling new invasions. What Hawai'i currently had, the report explained, was "a set of programs that are generally effective within their own jurisdictions but which, together, leave many gaps and leaks for pest entry and establishment." Only by stitching together all these different existing nets could Hawai'i hope to catch enough potential invaders to have any significant effect on the rate of new introductions.

Second, the report called for increased public support and involvement, a proposal that a few years later led to the Silent Invasion campaign. "Although public understanding of such apparent threats as snakes and other dangerous animals has increased through media exposure during the past year," the report's authors concluded, "the average citizen remains unaware of the magnitude of the alien species problem. Effective systems will require strong public support and participation, essentially making alien pest prevention and control a part of everyday life for people living in Hawaii." Increasing public awareness was an essential complement to improved interagency coordination because effective prevention and control systems are expensive and legislatures will fund control efforts only if a clear public mandate for alien species work exists.

These two tasks—improving interagency coordination and increasing public support and involvement—went from paper to action in early 1995 with the formation of an umbrella organization called the Coordinating Group on Alien Pest Species, or CGAPS. Membership in CGAPS is broad, with agencies from the Hawai'i Department of Agriculture to the U.S. Navy participating in quarterly meetings and providing staff support for projects. However, The Nature Conservancy served as both glue and catalyst for the partnership, particularly in its early years. By 1997, the Conservancy was dedicating about $100,000 in staff time annually to CGAPS, and most of the partnership's major projects had a strong Conservancy role. Even so, it was ultimately CGAPS's broad base—a total of fourteen major participating agencies and organizations—that gave it currency, and the combined contributions of all its members created a level of effectiveness that no one agency or organization could have reached alone.

The Silent Invasion Campaign is the principal public awareness effort of CGAPS. Part of the campaign is audience-specific, such as annual briefings for government officials and judges, training sessions for airline flight attendants, and workshops for teachers. These focused efforts complement CGAPS's more generalized public awareness strategies, which include placing informational displays in island airports, running ads in Hawaiian newspapers and tourist magazines, creating an educational video, and running public service announcements with island television and radio stations.

The centerpiece of the campaign is a twenty-eight-page full-size color booklet prepared for key leaders in government, business, and the media. Across its all-black cover glides a red snake, its tongue flickering and its body poised to strike. Below the snake are the words VENOMOUS SNAKES, KILLER BEES, TROPICAL DISEASES . . . HAWAII'S FUTURE IS AT STAKE. Open the cover and a huge tarantula extends its hairy legs across a yellow-and-red text that reads like a poster for a horror movie. The pages that follow feature piranhas, malaria-bearing mosquitoes, snakes, scorpions, fire ants—just about everything you'd rather not run into on the beach or in your backyard. "The potential impact of new pests on our health and lifestyle is immense," a

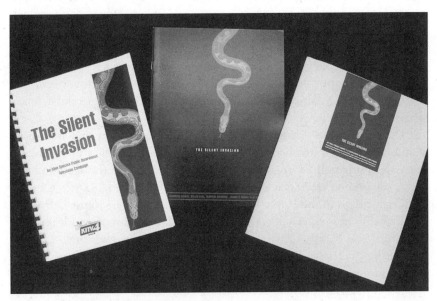

The Silent Invasion campaign's public awareness materials use vivid images to catch the attention of the public. (InfoGrafik)

headline warns. The brochure then shifts to industries at risk, and subsequent pages feature quotes from the chief economist of the Bank of Hawai'i, the president of the Hawai'i Visitors Bureau, and two Hawaiian farmers. It makes its point well: Alien species invasions matter to everyone.

It is only after health, safety, and the economy have been covered that threats to native species and forests are addressed. In a two-page spread, the brochure depicts the effects of mosquito-borne diseases on native bird populations, of feral pigs and introduced weeds on native forests, and of invasive ants on Maui's famous silversword plants. Overall, roughly equal space is devoted to biodiversity loss, watershed degradation, and erosion of native Hawaiian culture. Rare understory plants and native drosophilid flies are not headliners.

Making ecological issues almost incidental was a calculated move on the part of the CGAPS committee that put together the Silent Invasion booklet. "You are not going to get people's attention unless you focus on health and human safety and quality of life," explained Alan Holt, deputy director of the Hawai'i chapter of The Nature Conservancy. Only once the issue as a whole has begun to resonate with the public can you add agriculture and biodiversity. "If we had written this by ourselves three or four years ago," Alan said, "we would have led with the i'iwi getting bitten by malaria-ridden mosquitoes, and followed that with something about native biodiversity and how important it is. And we would have been talking to maybe twenty percent of our audience. There simply is not enough power behind biodiversity to carry the kind of solutions we need for this huge problem." By making the focus of the campaign dramatically broader than just rain forests and native species, CGAPS was able to overcome what had until that point been a seemingly insurmountable impediment to stemming the tide of invasions: the relative indifference of the majority of the population to biological conservation.

The campaign chose its headliner species very carefully. They had to be both a terrible ecological menace and an obvious threat to health, safety, and the economy. One logical choice was *Miconia calvescens*, a South American tree that is spreading aggressively across Hawai'i. Miconia has the potential to destroy the state's remaining native forests, as it has already done on the ecologically similar island of Tahiti. Under the leadership of Hawaii's governor, CGAPS is coordinating an ongoing publicity and control effort called Operation Miconia, aimed at eradicating the tree before it multiplies beyond control. While eradication is carried out by specific state and federal agencies or conservation organizations, CGAPS's role has been to raise awareness statewide so that a large segment of the population knows about miconia and reports it if they see it.

The other major focus species for the Silent Invasion Campaign is the brown tree snake. This voracious predator, native to Australia and Papua New Guinea, is not yet established in Hawai'i, but is present on the island of Guam. Since both military and commercial flights between Guam and Hawai'i are frequent, snakes could easily stow away in cargo and escape upon arrival in Hawai'i if exclusion and detection systems are not sufficiently rigorous. Hawai'i is currently a snake-free ecosystem, and the ecological and economic consequences of bringing such an obviously undesirable species into what both tourists and native forest birds consider "paradise" would be considerable. CGAPS has formed a committee of all the agencies with responsibility for keeping the brown tree snake out of Hawai'i to both improve the exclusion program and lobby more effectively for funding.

The first of these two campaigns will test the value of public awareness efforts as a component of prevention and control programs. Can flyers and television spots and door-to-door canvassing induce the public at large to care enough to make miconia eradication possible? This single-species campaign is built around the short-term goals of public awareness work: to convince the public to support funding to control the most immediate invasive species threats and to enlist the public as eyes-on-the-ground to locate new infestations.

The brown tree snake campaign is by nature a more long-term project. Eradication at least theoretically has an end date, but exclusion is forever. As a consequence, success in this campaign will indicate progress toward the long-term goals of public awareness work: to create a strong and lasting public constituency for a comprehensive system of prevention and control and to promote a new public ethic that actively discourages the importation of new pest species or their likely vectors.

While stopping miconia and the brown tree snake are in themselves fundamentally important goals, the most valuable outcome of these campaigns would be a statewide reconsideration of the true costs of a societal preference for the exotic over the native. Generating broad-based public support and involvement is only one of many elements of an effective system of prevention and control of nonnative species invasions, but it is perhaps the most elusive. If public awareness campaigns can create a strong constituency for this kind of work, everything else will become possible.

## The Making of a Crisis

The old coast road that heads north from Hilo was built before big bridges made straight roads possible in this part of the Big Island of Hawai'i. It drops down into one ravine after another, following the contours of the deeply incised landscape until it reaches an elevation where a modest bridge will do the job. Curves bring the road close to the many rivers of the Big Island's windward side, and the views are often breathtakingly beautiful. The Onomea scenic route, as it is called, is a must-see for visitors to the island, and my host, Nelson Ho, decided that I shouldn't be the exception to the rule.

Nelson parked his government truck a few feet before the first bridge, and we walked out to take a look at the forest that lined the boulder-strewn river below. As in almost any low-elevation forest in Hawai'i, almost all the trees along the river were nonnative. But Nelson's focus was entirely on one particular species of tree. He pointed down along the bank to where a clump of 3-foot-long green and purple leaves poked out above the canopy. "There it is," he said. "This is one of the oldest miconia infestations on the island."

And suddenly I was seeing them everywhere—*Miconia calvescens,* the most dangerously invasive tree ever to find its way to the Hawaiian Islands. The deeply veined glossy green leaves with bright purple undersides were strikingly beautiful and very easy to spot, once you knew what to look for.

This infestation was one of the most accessible on the Big Island, and calls had been coming in regularly to the miconia hotline from concerned

The large oval leaves are miconia trees that are taking over this forest along the Onomea scenic route on the Big Island. (Jason Van Driesche)

people reporting these trees to Nelson and his crew. As field supervisor for Operation Miconia on the Big Island, Nelson is ultimately responsible for eradicating this and all other stands of miconia on the island. For now, all he can do is thank everyone who calls and tell them that this patch is one on a very long list that needs treatment. Though he hopes to kill the miconia along this river within a year or two, all he can do for now is watch them grow.

Nelson first found out about miconia through the Sierra Club in the late 1980s. "It was an around-the-edges issue for me at first," he said. "I didn't realize how serious it was." By 1992, the magnitude of the threat had become clear, and Nelson began participating in Sierra Club workdays to control infestations as they were found. Within a few years, it became obvious that there was far more miconia on the Big Island than volunteers could handle, and in 1998 Nelson left a job as a lecturer in environmental studies at a local community college to work full time on miconia control. Though he tries to do other things on weekends, battling miconia has become his life.

Nelson is so dedicated because he knows that if miconia is allowed to multiply and spread unchecked, within a few decades it would probably obliterate a large portion of both the lowland alien forest and the remaining native upland forest in Hawai'i. Studies of miconia-invaded forests in Tahiti (whose ecological history and biota are comparable to that of Hawai'i) show that when miconia invades a site, it rapidly forms dense stands that suppress the growth of native understory plants and prevent the regeneration of native trees. Sixty years after its introduction to Tahiti, upward of two-thirds of the island has been invaded by miconia, and a third to a half of Tahiti's endemic plant species are directly threatened with extinction by the invader. Miconia also alters the water cycle in heavily infested areas, causing increased runoff and sedimentation of coral reefs. Its shallow roots and the lack of understory plants in pure miconia stands may even increase the incidence of landslides in Tahiti's steep, mountainous terrain. Miconia infestations in Hawai'i are only twenty-five years behind those in Tahiti. According to the late F. R. Fosberg, an eminent botanist who observed miconia infestations in Tahiti in the 1970s, miconia is "the one plant that could really destroy the native Hawaiian forests."

Why is miconia so much more invasive on Tahiti and in Hawai'i than almost every other alien plant species on these islands? In its native habitat in Central and South America, miconia is an early-successional tree adapted to finding and colonizing small gaps in the forest canopy. It produces up to 9 million seeds per tree every year. Its fruits are eaten and widely dispersed by birds, ensuring that at least some seeds are deposited in clearings with enough light for seedlings to survive. However, miconia never becomes par-ticularly abundant in its native range, both because there are a suite of nat-ural enemies in its native range that reduce its vigor and because other tree species in its native range are more shade-tolerant than miconia and soon overtop it, preventing miconia from perpetuating itself on any given site. This means that in its evolutionary home, miconia is a fugitive species, sprouting, growing, reproducing, and dying in rapid succession in the few years between the creation and the closing of a gap in the forest canopy.

But in Hawai'i (and on Tahiti as well), light levels in native forests are not low enough to keep miconia from germinating, for forests on these islands tend to be more open and short-statured than the dense, multilayered tropical rain forests of Central and South America. From miconia's point of view, Hawaiian forests are just one big clearing waiting to be filled. Since there appear to be no native tree species that are sufficiently shade-tolerant to suppress miconia, there is at least in theory nothing to stop it from filling every acre of suitable habitat on the islands. The ecological constraints that made miconia's growth habits necessary in its native habitat do not apply in Hawai'i. In its new ecological context, miconia is a plant out of control.

## Funding the Battle on the Big Island

Like most alien pest species, miconia's invasive tendencies went unnoticed for a number of years after its introduction. The tree was brought to the islands by a noted botanist in 1961 and planted as an ornamental at Wahi-awa Botanical Garden on the island of Oahu. Its unusually beautiful foliage must have attracted the immediate attention of other plant collectors and botanical gardens, for miconia was brought into the commercial nursery trade within a few years of its introduction to Oahu. Sometime over the next few decades, miconia was introduced to the Big Island, Maui, and Kaua'i as well. It wasn't until the early 1980s, however, that the conservation com-munity realized that the tree had become naturalized and was spreading rap-idly. By 1991, miconia had been recognized as a major threat to Hawaii's forests, and eradication efforts were begun soon thereafter on the Big Island, Oahu, Maui, and Kaua'i.

On the Big Island, it became clear very early that adding miconia control to the workload of overburdened public natural resource agencies and pri-vate conservation organizations was simply an inadequate response given the scale of the invasion. Existing agency budgets for nonnative species con-trol could not support the intensive multiyear campaign needed for eradica-tion. Particularly on the Big Island, the miconia infestation was so large that without an entirely new organizational structure for mobilizing support, wholesale destruction of native forest by miconia was inevitable. Generating public awareness was therefore an even more important component of miconia control on the Big Island than on the other islands.

To this end, scientists and conservationists on the Big Island put together an ad hoc group in 1995 called the Big Island Melastome Action Committee (BIMAC) to raise funds for eradication. (Melastomes are the plant family to which miconia and a number of other invasive species belong.) The committee's approach was modeled on a similar effort on Maui that had been successful in promoting the issue with county officials. BIMAC convinced the Big Island county council to provide initial funding for miconia control. Their efforts combined with the CGAPS-led statewide Operation Miconia campaign that was launched at about the same time ensured that public awareness about the threat of miconia was already relatively high when I visited the Big Island in 1997.

At least on the Big Island, though, this awareness had not yet translated into active public concern. Consequently, while the county council had agreed to fund limited control work on the Big Island, it was unwilling to spend enough to make eradication feasible. "Operation Miconia has generated major support for our activities in the field," Nelson explained, "but we haven't been able to mobilize that support in a way that produces long-term funding." Enough people on the Big Island knew about miconia to make it an issue, but too few people cared about it enough to make it a real priority.

A big part of Nelson's job was to do everything he could to push the issue higher on the county's agenda. "I'm being directed to put more time into securing the money we need," he said. Throughout the late 1990s, he met regularly with legislators to keep them informed of progress and talked with colleagues in state agencies that might be able to provide support for miconia control. But he knew that his case wouldn't carry much weight unless decision makers were hearing about miconia from their constituents. So he spent as much time as he could going door-to-door and talking with people who might have miconia in their backyards. Though personal contact is a powerful tool for generating support, the number of people Nelson could reach was small. Such on-the-ground public awareness work eventually had to be relegated to whatever he could do as part of his control activities, and television specials and newspaper articles had to do what Nelson did not have time to do personally.

By 1997, Nelson was spending most of his time either taking care of logistics for Operation Miconia or working with his crew controlling infestations. He took me to an abandoned canefield above Hilo where the crew was cut-

Nelson Ho is field coordinator of Operation Miconia on the Big Island, where thousands of acres of abandoned canefields are infested with miconia trees. (Jason Van Driesche)

ting miconia. There were trees 2 and 3 inches in diameter and 20 feet high every few feet for as far as the eye could see. A crew of ten could spend a month here and still not get every tree. Nelson had a crew of two. We got out of the truck at the limit of the area that had been treated, and Nelson pulled out his machete and waded in. He hacked through the bark all the way around the base of a large tree. Then he squirted herbicide into the cut, being careful to soak every inch of the tree's conductive tissue. "I've dreamt about miconia," he said, moving on to the next tree. "Everybody on the crew has had a miconia nightmare."

It's hard not to get discouraged, Nelson said, when you think of how many trees are out there, putting out who knows how many billions of seeds, and you don't have the resources even to get all the reproductive trees—let alone the seedlings they continue to produce. As of June 1999, Nelson and his crew had cleared about 7,600 infested acres, pulling 106,100 saplings and killing over 23,500 mature trees. They had also developed more efficient control methods that eliminated the need to cut the trees before treating them with herbicide. But even this massive effort and these improved techniques had at best only slowed the spread of miconia at the periphery of its range. Although some of the core infestations—such as the stands along the Onomea scenic route—have been treated by volunteers, many areas remain uncontrolled.

Nelson and his crew have a long and uncertain fight ahead of them. By his estimation, it will take at least ten years and $2 million to $3 million to do the job, and the situation is tenuous enough that even one year of lapsed funding would be disastrous. Even if efforts currently under way to develop an effective biocontrol program for miconia prove successful, an aggressive program of mechanical and chemical control will still be needed in the interim to contain the infestation and limit disruption of native ecosystems. If control is to be funded locally, continuous public pressure—active pressure, not just passive support—is crucial.

I gave Nelson a call in late 1999 to see if anything had changed since we last had talked about a year earlier. When I got him on the line, I could practically hear him grinning on the other end. "It's turned around!" he said. Realizing that the political and economic climate on the Big Island was such that the county government was not going to provide adequate funding, the Big Island Invasive Species Committee (BIMAC's successor) had shifted its focus to the state and federal level. With an offer of $150,000 in federal matching funds, they managed to convince the state legislature to appropriate $150,000 for miconia control on the Big Island. With $300,000 in hand, Nelson now stands a chance of overtaking the spread of miconia instead of just slowing it down. This money will fund a larger crew and helicopter time for locating remote infestations. It will also move biocontrol research ahead substantially—a crucial element in the long-term management of miconia. Most important, though, is the effect this much money will have on public perceptions of the effectiveness of the miconia campaign. Visible progress is what will bring in the next round of funding.

At present, public awareness and support still have a long way to go on the Big Island. Some local residents—like the man who told me about how he had made his daughter-in-law rip up a miconia tree she had planted in her yard ("That plant is bad news!" he had scolded her)—are already actively concerned about miconia. He is probably still more the exception than the rule on the Big Island, though. Until there are a lot more people like him on the Big Island, miconia control there will have to rely on public awareness at the state and national levels.

## Local Residents as Allies on Maui

The miconia situation is for several reasons significantly more manageable on Maui than on the Big Island. First, miconia infestations on Maui are much less extensive than on the Big Island. The single largest infestation is a 2,000-acre patch on state land above Hana on the east end of the island. Second, the fact that this site is entirely in public conservation ownership greatly simplifies control efforts, because it eliminates the logistical headache of doing control work on a patchwork of privately held parcels. (While all the other known infestations on Maui are on private land, there are only a dozen or so sites, and none is larger than a few acres.) Finally, miconia has been on Maui a decade less than on the Big Island, and this ten-year difference appears to have been critical.

But there are other factors besides the scale of the infestation that have made eradication more likely on Maui than on the Big Island. High on the list is the active support the Maui community has given to miconia control. Particularly in the Hana area, where the control effort has provided the community with much-needed jobs and a sense of accomplishment, local residents are quick to speak out in favor of miconia work. As eyes in the forest and voices at county board meetings, Maui residents have come out strongly in favor of immediate and decisive action to eradicate miconia from the island. On Maui, citizens are much more an ally than a passive backdrop.

I called Pat Bily on Maui as soon as I flew in to Honolulu. Pat is the invasive plant specialist for The Nature Conservancy's Maui field office and spends most of his time in the forest. After a few days of trading messages, I finally managed to catch him in his office. "I have some good news for you," he said. A call had just come in to the Maui miconia hotline from a landowner reporting a tree on his property, which meant that I would have a chance to participate in scouting and perhaps even control work with Pat when I was on Maui. I arranged to head out into the field with him a few days later.

The property with the new infestation was on a hill overlooking the ocean, about 15 miles out the Hana road along windward East Maui. It was a good thing Pat already knew the landowner, or we never would have found the narrow, brush-choked driveway that ran the quarter-mile from the road to the house. Two men came down from the house as we pulled up, and Pat shouted a greeting out the window of the truck. "The tall one is Wolfgang. He's the owner," Pat explained as he turned off the engine. "The other guy is Eddie. He's a local guy—a hunter. You'll like him."

Pat got out of the truck and shook hands with both men. After introductions were made, we headed back down the driveway to the place where the miconia had been sighted. As we walked, the three of them talked about local politics and the weather and what was going on over the weekend. Pat asked Eddie how his family was doing, and Eddie told us about the

Miconia control crew leader Darrell Aquino and his team have put in thousands of hours cutting and herbiciding miconia on east Maui. (Hawai'i Division of Forestry and Wildlife)

pig he'd killed last time he went hunting. About a hundred yards below the house, Wolfgang stopped and pointed down into the gully to our right. "There it is," he said. "Eddie noticed it the other day when he was cutting back the brush along the driveway." A crown of purple and green leaves was just beginning to emerge above the canopy about a hundred yards downslope. If Eddie hadn't seen the tree and called it in, Pat would never have known it was here.

Eddie led the way with his machete, cutting a path up the gully toward the miconia tree. He is nearly sixty, wiry and tough, with one hand that is permanently curved to fit a machete handle and one that knows how to stay out of the way of the blade. He spends a lot of time in these forests hunting pigs or just rambling around and keeps an eye out for miconia wherever he goes. Earlier in the day when we had turned into the driveway, Pat had noticed a miconia seedling hanging on a fence with its roots pointed skyward. When he asked Eddie if he had put that seedling there, Eddie just grinned. Then he told us he found it growing in a nearby pasture and wanted to make sure it didn't have a chance to reroot.

Pat and Eddie met at a county budget hearing around 1995, where both of them showed up to testify in favor of county funding for miconia control. Eddie had learned about miconia from his brother, an accomplished amateur botanist. Pat and Eddie started talking after the meeting, and Eddie has been a volunteer with The Nature Conservancy ever since. "Hunters like Eddie are a critical element," Pat said. They know the mountains better than anyone else and cover far more ground than Pat ever could. Along with helicopter reconnaissance, hunters are Pat's principal tool for finding new satellite infestations. The biggest difference is that helicopter time runs at least $600 an hour, but once hunters learn a little about miconia, they help out for free.

It took us about twenty minutes to work our way up to the tree we'd seen from the driveway. Contrary to what Pat had expected, there were no seedlings along the stream. "That's a good sign," he said. But he stopped short as the tree came into view. "Hooo, boy," he exclaimed. "That's one monster tree." It was at least 25 feet tall, and the upper branches were loaded with clusters of mature fruits. Pat and Eddie took turns with the machete, and in five minutes the tree was on the ground. Branch by branch, we cut off the fruits and triple-bagged them in plastic garbage sacks. Pat would take them back to town and incinerate them to make sure the seeds wouldn't spread.

As we worked, Pat pointed to the dried-out fruit stalks from last year that were still attached to the branches. "This tree has already put out millions of seeds," he said, shaking his head. Soon there will be thousands, perhaps tens of thousands of seedlings growing around the herbicide-painted stump of the parent tree. Someone will have to come back here every year for five years or more until the seed bank is exhausted. As we left, Eddie almost absent-mindedly pulled up a tiny miconia seedling that he'd spotted growing on the bank of the creek. He'll be keeping an eye on this place.

Though Pat has local contacts across East Maui who help him find new infestations and keep an eye on the ones he already knows about, none of his other contacts is as dedicated as Eddie or understands the threat like he does. "Eddie has a sense of the innate value of these forests," Pat told me on the drive home. He has what native Hawaiians call a love of *aina*—the Hawaiian word for "the land." He sees miconia as a threat to the *aina*, so he fights against it as few other people do.

Still, there are enough people on Maui who value what is unique about Hawai'i that when county budget hearings come around, invasive species are generally on the agenda. It was a combination of scientific testimony and local support that convinced Maui County to provide funding for miconia control. And it is in part the environmental community that elects people like Avery Chumbley and David Morihara—both strong supporters of alien species control—to the state legislature. While people like Eddie Oliveira do not by any means constitute a majority of Maui residents, there are enough like him that Maui's political climate is distinctly "greener" than that of most of the other islands. Perhaps more than any other factor, it is this kind of generalized local awareness and support that has made nonnative species battles more winnable on Maui than almost anywhere else in the state.

## "Whole Island, Whole Public"

The kind of broad-based environmental awareness that characterizes Maui does not just arise spontaneously. Conservation organizations and agencies on Maui have proven particularly adept at planning, securing funds for, and carrying out alien species control efforts on a broad scale—one that includes public awareness work as well as on-the-ground management. "Maui is really together in terms of alien species work," explained biologist Pat Conant, a Big Island resident active in alien species issues. "The rest of us are just sleeping. Maui is the model."

As such, alien species work on Maui in the 1990s and beyond is the first full-scale trial of the approach CGAPS has advocated statewide. For whatever reason, the conservation community on Maui is blessed with a large number of remarkably dedicated and effective individuals—people who, as Maui forester Bob Hobdy put it, are willing to stand up and say, "I will do it." The miconia campaign began on Maui when a scientist named Betsy Gagné—who had just come back from a visit to Tahiti—discovered a single miconia tree along the coast road on the island's east end in late 1988. By early 1991, Gagné and others realized the true magnitude of the threat and began spreading the word. A series of volunteer work parties were organized to eradicate the invader, and within two years, over 20,000 miconia plants were removed. By mid-1991, front-page articles in the local *Maui News* and the statewide *Honolulu Star-Bulletin* had brought the issue to the attention of the general public, and more infestations were located along the east Maui coast following the distribution of an eye-catching "wanted" poster on miconia. In response to the ascending spiral of infestations, a group of Maui conservation professionals decided in August 1991 that pooling their expertise and their resources was the best way to deal with miconia and related plant pests, and the Melastome Action Committee (MAC) was born.

The committee immediately began a coordinated program of field scouting, research, and control. Lloyd Loope, a research scientist stationed at Haleakala National Park and a leader in the fight to get miconia recognized as a major threat to Hawaii's native forests, convinced his superiors at the National Park Service (NPS) that park funds would be better spent eradicating miconia from the lowlands immediately than fighting it later after it moved upslope and invaded the park. NPS-sponsored efforts were directed in large part toward scouting and control of populations along the East Maui coast and research on the biology of miconia.

The battle escalated dramatically in September 1993, when Bob Hobdy discovered the island's largest infestation high on the slopes above Hana on Maui's east end. There were miconia trees scattered over hundreds of acres of jumbled lava, and the trees had already been setting seed for a decade or more. Once the scale of this infestation became clear, the conservation community realized that miconia was not just a concern, but an emergency.

In June of 1994, Loope sent his co-worker Art Medeiros to Tahiti to get some perspective on just what MAC was up against. When Medeiros came back with dramatic photographs of entire mountainsides heavily dominated by miconia and dense miconia stands entirely devoid of understory vegetation, the committee knew it had something powerful. Using these photos, MAC made a presentation to the county council about the threat of miconia to Maui's forests. Loope described the reaction of the council members, many of whom were from agricultural or rural backgrounds. "A hush went through the council when they saw the photos from Tahiti," he recalled. The council members knew how important watershed protection was in an island state and recognized the threat of miconia not only to native ecosystems, but also to the island's water supply. "Right there," Loope said, "the council committed to funding miconia control."

Over the next three years, the county contributed $264,000 to the effort—well more than a third of the total funding from all sources for Maui-specific miconia work. MAC clearly had done its job, for the local community was taking the initiative to ensure that government at the local as well as the state and national levels contributed to the eradication effort. The committee continued to meet regularly to coordinate the campaign, for even with adequate funding, getting rid of miconia was an enormous task. As of late 1999, the core miconia population at Hana and all known peripheral infestations had been treated, and follow-up to remove new seedlings was ongoing. Eradication is by no means ensured, but it is a real possibility.

Once control work was under way, the committee decided to take a longer view of the invasive species problem in general. The miconia crisis

In Tahiti, where pure miconia stands like this one have eliminated native forest across much of the island, the shallow root systems of miconia trees may increase the risk of landslides. (Paul T. Holthus)

had shown MAC's members how expensive a major eradication campaign could be—and by contrast how cost-effective prevention would have been. With public awareness and concern about invasive species at an all-time high, MAC's members felt that the community was ready to do more than simply react to one crisis after another.

In December 1997, the Melastome Action Committee reconstituted itself as the Maui Invasive Species Committee (MISC) in order to address a wider range of alien species threats. MISC spent much of 1998 developing a strategy, including a list of priority species for control, eradication, or monitoring. The emphasis was on those species that were not yet widespread, but had the potential to become major pests. A proposal submitted in November 1998 to the National Fish and Wildlife Foundation brought in $206,000 in federal funds, and the state legislature, the county council, the county water board, and The Nature Conservancy contributed an additional

$540,000 for MISC's operations. Eight years of constituency building had paid off.

MISC is using the money to hire a coordinator, a public awareness specialist, and a permanent, full-time crew to serve as a rapid response team. These new positions will relieve Bob Hobdy and Pat Bily and all the other members of MISC of the need to fight brushfires and allow them to focus their energies on planning, public awareness work, research, education, and other long-term endeavors. "We're really going all out to try to get materials into the schools," Loope said. The long-term goal is to find ways to continue to expand the constituency for alien species control into the community as a whole.

In effect, the committee has begun to move beyond developing an effective response capacity to building a broad public constituency for policies that would close the door to new invasions. In early 1999, Haleakala National Park—an important member of MISC—submitted a grant proposal to the Biological Resources Division of the U.S. Geological Survey for a project designed to prevent alien species establishment through outreach. The project would focus on developing a comprehensive "map" of all the pathways by which new invaders might be brought in, testing methodologies for early detection and rapid assessment of new introductions, and mapping and monitoring nonnative species present on Maui that might become invasive. Two project staffers would be responsible for establishing contacts with those industries—agriculture, landscaping, pet stores, and so on—that are responsible for most new introductions. They would also do a great deal of field work in the populated areas where many potential invaders of natural areas are already established. Consequently, they would spend much of their time raising the issue of alien pest species in one way or another with a wide range of businesses and groups. While the overall purpose of this project is to gather information essential to the creation of a comprehensive and vastly improved system of proactive response to nonnative species threats, its outcome would do just as much to spread the word as it would to gather data. On Maui, the long haul has begun, and outreach to the community will be a critical and ongoing part of the task.

## Putting Public Awareness to the Test

The first real test of the power of public awareness to influence invasive species policy on Maui began to unfold in the late 1980s when citizens started negotiating with state and federal officials over a proposed expansion of Maui's regional airport that would open it to direct flights from Asia and increased traffic from the U.S. mainland. Currently, most passengers bound for Maui are routed through Honolulu. Conservationists fear that internationalization of the airport would create an open conduit between sources of invaders and the island of Maui, thus increasing the rate of nonnative species introductions to the island. The invasive species problems Maui currently faces pale in comparison with what could come with an airport that plugs the island directly into the global transportation system.

This kind of issue—one where the invaders are as yet only hypothetical but the funds and the dedication needed to keep them out are quite substantial—serves as an effective measure of how far public concern about invasive species extends and of how successful the Silent Invasion campaign and the Maui Invasive Species Committee have been at strengthening that concern and drawing it into public forums where such decisions are made. Is public awareness strong enough that the state will be willing to make substantial changes in the way it interacts with the outside world in order to prevent invasions, or is official concern limited to throwing money at obvious crises? The competitiveness of the island's tourism base and its ability to export perishable agricultural commodities are seen to be at stake, for the Kona side of the Big Island expanded its airport in the mid-1990s and Maui's farmers and hotel owners don't want to be left behind. The island's two major industries—agriculture and tourism—have an immediate financial interest in the outcome of the debate that tends to push a "progress-at-all-costs" mentality, particularly on the part of the state and federal agencies mandated to maintain the island's transportation infrastructure.

One of the most well-publicized threats of internationalization is the increased likelihood that the brown tree snake would be introduced to Maui on direct flights from snake-infested areas. Maui is home to many species found nowhere else in the world, including a variety of forest birds. On the South Pacific island of Guam, where the brown tree snake was accidentally introduced from Australia or Papua New Guinea sometime after World War II, predation by snakes has caused the extinction of ten of Guam's thirteen bird species. There are now up to 12,000 snakes per square mile on Guam—the highest snake population density in the world. This invasive reptile had been present on Guam for years before anyone realized it was there, for no one paid any attention to nonnative species introductions in the 1950s. And it was only after birds had virtually disappeared from Guam—only after the forests were silent—that a group of researchers finally established the link between the snakes and the loss of wildlife on the island. The snakes on Guam also attack people (especially babies) while they are sleeping and cause power outages every few days by climbing on utility poles. They now threaten to invade other Pacific islands and even the U.S. mainland by hitchhiking on flights out of Guam.

Hawai'i is equally vulnerable to the impacts of the brown tree snake and has even more to lose. Lloyd Loope estimates that if the brown tree snake were introduced to Maui, it would spread across the island within about forty years. And once the snake was on one island, it would be extremely difficult to keep it from hitchhiking on interisland flights and invading the Big Island, Oahu, and the rest of the state. There is no known way to eradicate an established population of brown tree snakes. This is why keeping the snake out of Hawai'i is such a high priority: If a pregnant female escapes into the wild, it's all over.

But the Silent Invasion Campaign has done its work well. Television spots about what the brown tree snake has done to Guam have heightened public awareness about this potential invader in Hawai'i. Almost everyone in the state has at least heard of the brown tree snake. Even President Clinton, who visited Hawai'i in 1996, had been so thoroughly inundated with press releases and editorials about the brown tree snake that at a press conference in Honolulu, he held up a clipping from a local newspaper that editorialized against the snake and he pledged to help. Shortly thereafter, Congress approved $1.6 million for brown tree snake research and control—nearly three times the amount that had been allocated in previous years.

While the brown tree snake remains a centerpiece of the public awareness campaign, concern about the impacts of all invasive nonnative species is growing statewide. Particularly on Maui, an attitude of "No more invaders, period" has begun to take hold. This broad public awareness and support for alien species prevention and control emerged with the battle against miconia, but has expanded further still through the debate over internationalization of the Maui airport. In newspaper articles and county council resolutions, keeping out invasive pests has become a major public issue across Maui.

You wouldn't know alien pest species were an issue, though, from the amount of attention they received in the early stages of the airport planning process. When the Hawai'i Department of Transportation (HDOT) began planning for expansion of the Maui airport in the early 1980s, there was almost no mention of alien species impacts in its planning documents. In 1988, when the revised plan included provisions that would have allowed for direct international flights, conservationists began to voice their concerns.

Haleakala National Park raised the issue of possible impacts of expansion on the island's native ecosystems and began to try to influence the planning process and cultivate support across Maui. In a statement to the Maui County Council in January 1991, Park Superintendent Don Reeser said, "We only comment on proposed development we believe would be detrimental to the resources of Haleakala National Park. The internationalization of the Kahului Airport is one of these projects ... [because it] would result in a significant increase in the number of alien weeds and insects on Maui." Consistent with the park's concern was the fact that the state's environmental assessment for the project was thrown out by the state court because of an inadequate analysis of the ecological consequences of internationalization. In response to the comments of the park representatives and other concerned citizens, many of whom were against expanding the airport, the council subsequently adopted a resolution opposing internationalization.

Over the next couple of years, the project went through several rounds of planning and litigation as state and federal authorities attempted to prepare a document that met the stipulations set by the state court. Eventually, the Federal Aviation Administration (FAA) took over as the lead agency in the preparation of a joint state and federal Environmental Impact Statement (EIS). A few months before this process began, the Park Service requested that it be included as a cooperating agency in the drafting process so as to ensure that the threat of alien species introductions to the island was adequately addressed in the EIS. The FAA denied its request, asserting that the park was too far from the airport and that, in any case, impacts on endangered native species were not likely to become an issue in the EIS. The threat of alien species introductions was nowhere mentioned in the FAA's response to the park.

What the Park Service and the Maui conservation community were confronted with was a federal agency that is mandated to expand airport capacity "to the maximum extent feasible." It is unlikely that the FAA has any biologists on staff, for conservation of native ecosystems and preventing nonnative species invasions are not in its mission. It is concerned solely with making air travel safer and more efficient. Everything else is secondary.

Over the next two years, the Park Service worked both behind the scenes and in the public arena to keep the pressure up on the FAA to consider alien species impacts. When the Draft EIS was released in March 1996, though, it became clear that the FAA had not been listening. A May 7 article in the *Maui News* entitled "Haleakala Chief Says Airport EIS Has Gaping Hole" set the stage for a public hearing on the Draft EIS that was scheduled for a few days later. Following the hearing, another article reported that, of the sixty-four people who testified, nearly nine out of ten "blasted it as being 'totally inadequate.'" The article also noted that "by far, the most frequently cited complaint about the EIS was its treatment of alien species introductions and their potential effects on Haleakala National Park." The Maui community had clearly sided with the park in its efforts to prevent new non-native species introductions to the island.

The FAA was taken aback by the vehemence of the criticism leveled against the Draft EIS and recognized that opposition was conceivably strong enough to derail the expansion project entirely. In response, the FAA called a meeting in June 1996 with representatives of the National Park Service, the U.S. Fish and Wildlife Service, and the Hawai'i Department of Transportation and offered to prepare a Biological Assessment (BA) as part of the final document. Haleakala National Park was invited to participate in the process, as were prominent scientists from other concerned agencies and organizations. The primary purpose of the BA would be to assess the threat of alien species introductions via the airport and to outline mitigation measures designed to prevent new invasions. Lloyd Loope noted the apparent change of heart in the FAA, writing in a memo for park files, "Since they are aware of apparently strong feelings by the public on Maui on the subject of alien species introductions, they may be willing to support substantial, yet reasonable, mitigating measures."

But as the preparation of the BA progressed, it became clear that, as Loope later put it, the FAA "was merely 'going through the motions.' There seemed to be little interest in incorporating adequate alien species measures into the project." His assessment was proven accurate when the Final EIS came out in late 1997, in which the provisions for keeping out alien species appeared little improved. Particularly telling was the statement that began the section of the EIS that described proposed mitigation measures: "There are no significant impacts to endangered and threatened species or to native species due to the Proposed Project, and therefore no mitigation is necessary." The FAA had prepared a mitigation plan for something it did not believe needed mitigating, and their lack of concern showed in the document's lack of substance.

At that point, the National Park Service had no choice but to prepare a referral to the President's Council on Environmental Quality, a high-level body charged with arbitrating environmentally related disputes between federal agencies. The Park Service contended that the FAA's proposed actions—including public awareness programs, increased inspection, and other measures—were unacceptably vague and largely without substance. The Park Service's conclusion was unequivocal: "From the standpoint of the NPS, the mitigation measures agreed to by the FAA will be largely ineffective. They are non-contractual and [are] unlikely to intercept alien species." The proposed measures provided only a marginal increase in prevention while allowing a large increase in risk. The NPS position was supported by letters from dozens of Maui residents and other concerned individuals, many of whom went to great lengths to describe their misgivings about the way the proposed airport expansion had been handled.

Several months later, the Council on Environmental Quality passed the matter on to a group of high-level representatives of the Department of the Interior, the U.S. Department of Agriculture, the FAA, and other federal agencies who had been chosen to resolve the debate. The product of their negotiations was an Alien Species Action Plan (ASAP), which was included in the FAA's August 1998 Record of Decision approving the expansion. Keeping the airport a regional facility—the option endorsed by the Maui County Council only a few years before—was no longer even on the table.

The stated goal of the plan is "preventing the introduction of alien species into Maui via the Kahului Airport to the greatest extent possible." However, the plan still relies on voluntary participation by airlines for many of its public awareness measures. It still uses phrases such as "will consider" and "if feasible." Most important, the last paragraph begins with the critical caveat: "Subject to the availability of funding . . ." This kind of language would never be found in a document outlining measures to deter terrorist attacks. Preventing alien species impacts may be a top priority for the Maui community, but the mainland-based federal officials who worked out the plan clearly see the issue as a hurdle, not a bottom line.

At the state and national levels, awareness and concern are not nearly as strong as they are on Maui, and expansion may win out over protection. As far as the FAA is concerned, cutting two hours off the flight to Maui is more important than preventing the likely devastation of the island's native ecosystems, its agricultural base, and its way of life.

On the local level, though, public awareness work related to nonnative species impacts has been remarkably effective on Maui. A large segment of the local community stood in strong opposition to internationalization of the airport, largely out of a desire to protect the island from the impacts of invasive nonnative species. Though the future of the Kahului Airport expansion is still very much up in the air pending the outcome of a National Parks and Conservation Association lawsuit, one thing is certain regardless of the outcome: Invasive species impacts will be a major political issue on Maui from here on out.

When the CGAPS coalition began work in the mid-1990s, the task it had set out for itself was enormous: to close the gaps in the alien species prevention and control system through better interagency coordination and expanded public awareness. It took on three specific challenges in its first few years—containing and eradicating miconia, keeping out the brown tree snake, and raising general public awareness about these and other alien species issues statewide—with the intention of making them all what The Nature Conservancy's Alan Holt called "home runs."

By any measure, CGAPS had by 1999 catalyzed tremendous progress in all three areas. Miconia has been eliminated from Kaua'i and is on its way out on Oahu and (one hopes) on Maui as well. Even on the Big Island, control efforts have advanced to the point where eradication is beginning to look like a possibility. Brown tree snake preparedness is, according to Alan Holt, "light years ahead of where it was." And public awareness of alien species issues in general is far greater than it was only a decade ago.

While support for alien species control and exclusion has increased statewide, progress is by no means even from one island to another. The kind of active public awareness and broad interagency coordination that CGAPS has been advocating as the foundation of alien species prevention and control is firmly established so far only on Maui, where concern about nonnative species impacts on the island's ecology and the economy was already strong among local residents even before CGAPS was formed. Although other islands have begun to follow Maui's model—Oahu has established a watershed partnership effort similar to the one that has for the last decade anchored conservation efforts on East Maui, and the Big Island recently formed a Big Island Invasive Species Committee that is modeled on MISC—alien pest species have yet to become a core issue statewide.

But everything is relative. Though the other islands lag behind Maui, awareness about invasive nonnative species is much greater throughout Hawai'i than almost anywhere on the mainland, and meaningful change in state policy is much more likely in Hawai'i than anywhere else. Mark White of The Nature Conservancy's Maui office is confident that within about five years, nonnative species prevention and control will be given consistent regulatory priority statewide, will feature prominently in any major policy debate, and will be adequately funded even when the economy takes a turn for the worse. And really, this is the point of everything CGAPS has been trying to do from the start: to get lots of people to see alien species as an issue as basic as schools or highways or police protection.

*Ua mau ke ea o ka aina i ka pono.* "The life of the land is perpetuated in right-eousness." This is the Hawai'i state motto, but for the growing number of people who are fighting to protect the things that make Hawai'i unique, it means something more. This phrase implies a duty to protect the *aina*—the land—as the foundation of everything that is Hawai'i. Public awareness campaigns are fundamentally about bringing the state motto to life, for when these words move off the state seal and into the way people live, Hawai'i will have remade itself as a place once again native.

CHAPTER 14

# Going Local

## *Personal Actions for a Native Planet*

*Palimpsest.*

This long-forgotten word describes the practice of reusing parchments by scraping the ink off a document that is no longer needed and writing new text in its place. In this practice are reflected the practical limitations of an age in which manufactured goods were in short supply. Before the dawn of the industrial age, writing materials were a scarce resource used again and again, each year carrying the ideas of a new set of thinkers.

Today civilization rewrites landscapes like parchment, inscribing what it wants from the land on its hills and streams and forests. With 6 billion people spread out across the earth's surface, land has become the ultimate limiting resource. The human story is to varying degrees written almost everywhere, and the few places where civilization's mark is still faint are disappearing fast. For most people, the only option that remains is to write on the land again and again, removing marks that are erasable and writing around those that have proven permanent. Unfortunately, some of the most long-lasting marks on the land—the ones that prove most difficult to write around—are the marks left by invasive species.

Much of this book has been concerned with putting down on paper a story that already exists, but had so far been told only in fragments. The intent has been to describe the impacts of introductions already made and point toward the likely consequences of a few more decades of more of the same. The book closes with an antidote. While some part of the damage people have done will last as long as the human race, there is much that can still be fixed. And if enough people begin to write a positive future on the land, it might still be left much improved.

Invasive species problems are often so huge as to inspire only despair.

How can one person possibly make any difference when one shortsighted importation can do more damage than anyone can undo in a dozen lifetimes? While it is true that many issues related to nonnative species have to be resolved at the societal level, there are many things that an individual can do to make any situation a little better. What follow are a series of "problems" and a few ideas of what people can do—both individually and working together—to be a part of a solution.

## Individual Actions

### *At Home*

PROBLEM: Most landscaping is done with nonnative plants. While many of the species used are perfectly innocuous—tulips are not about to become invasive—others are among the worst invaders of natural areas. Many nurseries not only offer known invaders for sale, but regularly introduce new (and therefore potentially invasive) species into the trade as well. All the characteristics that plant hunters look for in a new ornamental—hardiness, adaptability, prolific seed production, and so on—are also the characteristics of a perfect invader. As a consequence, natural areas in and around cities have to deal with more and more invasive species threats every year.

WHAT TO DO: Some nurseries and mail-order seed companies are beginning to offer native plants for sale and would probably be happy to discuss native and noninvasive exotic landscaping options with customers. Buy plants and seeds from reputable sources, though—some native plant dealers have been known to dig up plants or harvest seeds from wild populations, which can be devastating for rare or restricted-range species. And make sure that when a plant is sold as native that it is actually a species found in the local area. The term "native" is sometimes defined very broadly as "native to the United States," and many species native to one part of the country can be invasive in another. Local information is essential, and a good nursery that sells native species should be able to help. If not, the exotic pest plant councils and native plant societies that now exist in many states generally have lists of invasive species to avoid and suggestions for native and noninvasive introduced plants appropriate for various landscaping needs.

PROBLEM: "Exotic" pets—spiders, lizards, snakes, birds, cockroaches, fish, and so on—are becoming increasingly popular. In most states (except Florida and a few others), there are currently few restrictions on the sale of any species as pets so long as they don't eat plants. Different species pose different levels of risk. While certain species can under some circumstances be considered safe, the industry as a whole has introduced many nonnative species, a portion of which have proven invasive. More than thirty species of tropical fish have become established in Florida because floods washed them out of breeding ponds and into nearby rivers and lakes or because people released unwanted fish into streams or ponds. Buying exotic pets supports an industry that is already responsible for numerous introductions and encourages the importation of yet more new species.

WHAT TO DO: The easiest solution is to stick with the familiar. There are already plenty of dogs and cats and gerbils all over the United States, so buying a more traditional pet won't introduce a new threat to native ecosystems.

PROBLEM: Even the most familiar pets can do tremendous damage if they are released or allowed to roam free in a vulnerable environment. In 1989, a well-meaning resident of the island of Maui released six unwanted pet rabbits inside Haleakala National Park—thinking perhaps that a park was the perfect place for bunnies to be happy and free. But their progeny multiplied rapidly in Maui's predator-free environment, and the Park Service discovered the population just barely in time to eradicate it before it spread out across the park and destroyed the native vegetation. Cats can do similar damage to native songbird populations if they are allowed to roam free. Just because a species is domesticated doesn't mean it can't become invasive.

WHAT TO DO: Keep pets inside or on a leash, and never release a pet into the wild. It sounds cruel to deprive a pet of its liberty, but the unseen cruelty to native species—be they birds or plants or anything else—can be much greater.

PROBLEM: Imported natural products can harbor unseen pests. Everything from lumber to flowers to wood carvings can have insects or fungi or other potential pests on them or in them, many of which are often difficult or impossible to see. Fumigation alleviates the problem to a certain extent, but there are many kinds of pests (for instance, wood-boring insects) that fumigation may not kill, and not everything gets fumigated. (And in any case, one of the most common fumigating agents—methyl bromide—is extremely toxic and a very significant contributor to ozone depletion.) A second species of Asian longhorned beetle that attacks softwoods was recently discovered in the wooden "trunks" of artificial Christmas trees from China. Native bromeliads in Florida are under attack from a species of weevil that came in on bromeliads imported from the tropics. Just about anything that was once growing and that hasn't been heat treated or otherwise thoroughly disinfected can serve as a vector for pests.

WHAT TO DO: This is another good reason to buy local—or at least to buy American. The globalization of commerce means that a bouquet of flowers is as likely to be from Colombia as from North Carolina, and a two-by-four will as often be from Chile as from Oregon. It often takes a little detective work to find products from one's own biome, but it's worth the effort. Of course, there are many processed imported products—a bottle of wine, for instance, or a package of cheese—that pose no particular risk of carrying invasive pests, and there is no reason to avoid such items. Where the line needs to be drawn is at the importation of unprocessed natural products that are related to—and therefore can carry pests that would attack—a native species.

PROBLEM: Globalization has fueled the development of many industries in developing countries that are geared specifically toward markets in the industrialized world. Some of these industries rely on large-scale introductions of nonnative species, many of which have caused enormous ecological problems. For instance, a large proportion of the shrimp consumed in the United States is raised on shrimp farms in tropical countries around the world. When shrimp breeding stock is shipped from one farm to another, it often carries pathogens that can infect native shrimp populations in the area where shrimp are being farmed. (Fish farms may vector diseases as well.) Since shrimp and other small crustaceans often form the basis of marine food chains, the introduction of a particularly virulent disease could cause the collapse of entire marine ecosystems.

WHAT TO DO: Eat U.S.–produced shrimp and fish. Until foreign producers can be certified as environmentally friendly, they present too big a risk to the integrity of marine ecosystems.

## On the Road

PROBLEM: When weeds make their way into remote natural areas, they rarely get there on their own. Invasive plants along trails often come in on the boots and gear of backcountry travelers or in the droppings of pack animals. Areas with many trails and heavy human use may be at risk from such invaders.

WHAT TO DO: Cleaning all mud and burrs off equipment after every trip solves the problem for hikers. People who use pack animals need to take the extra precaution of buying feed that is certified free of weed seeds.

Some companies that run pack trips have begun using weed-free feed, in part because many natural areas (for instance, most national forests) now require it. Demand certification when you book a trip.

PROBLEM: Off-road vehicles transport weed seeds as well and are even more efficient at spreading invasive plants than are boots and horses. Seeds of all kinds are carried in wheelwells and get stuck in the mud caked on sidewalls and fenders, and the disturbance caused by knobby tires creates an ideal seedbed for many weeds. It is sometimes possible to see where an off-road vehicle has passed through a meadow just by looking for the stripe of invasive plants. Boats towed between lakes or ponds can tranport invasive species like Eurasian watermilfoil and zebra mussels directly to new water bodies as well.

WHAT TO DO: Stick to established roads and trails. There is no way to go off-roading without spreading invasive plants and tearing up the land. Clean your boat's prop and hull and drain the well before you put it again in another body of water.

PROBLEM: Cheap airfares have brought international travel within the reach of a large segment of the population. With many travelers taking advantage of opportunities to see the world, more people may be tempted to bring back interesting but biologically dangerous items such as colorful seeds, plant cuttings, odd pets, and so on.

WHAT TO DO: Photos are completely risk-free. Shells and rocks are probably a pretty safe bet. It's better to leave the rest—fruit, pieces of wood, seeds, flowers, and everything else that's alive or could have live creatures in it—in place for the next person to enjoy.

## Taking Action Together

### In the Local Area

PROBLEM: Many garden clubs still promote an exotics-centered approach to gardening. This leaves their members without a model of how to design and maintain an attractive native-friendly garden.

WHAT TO DO: There are some garden clubs that already work with native plants, but few employ an integrated native-friendly approach—that is, one that uses native species and judiciously selected noninvasive exotics and actively avoids any species that might be invasive. Clubs will no doubt welcome new ideas from members, particularly if a member is willing to teach a class or offer a tour. The best way to promote this kind of gardening is to show how well it works. There is nothing more inspiring to someone considering the possibility of making their garden native-friendly than seeing a beautiful garden that has already made the transition.

PROBLEM: County and municipal governments and agencies tend to use mostly nonnative species for landscaping and ground covers. Highway departments are among the worst culprits. Regraded highway rights-of-way and new road corridors are almost invariably seeded with nonnative ground covers—often invasive grasses such as smooth brome—because these are generally the kinds of species whose seeds are cheapest and most readily available. Municipal parks tend to create manicured exotics-based landscapes, maintaining lawns and ornamental shrubs even in areas where a low-maintenance mix of native species would be appropriate.

WHAT TO DO: Cities and other developed areas are exciting places to do natural community restoration projects. Most urban parks contain at least remnants of native vegetation, and parks departments are generally quite happy to enlist the help of neighborhood groups for restoration projects. It is important to seek professional ecological advice and planning assistance, but the most important ingredients in a restoration project are enthusiasm and willing hands. Persuading the highway department to revegetate with native species is often a little more difficult, since there is generally no mechanism for involving volunteers in highway work, and getting the department to change its own practices may not be easy. A first step might be to provide the head of the highway department with a list of particularly invasive species to avoid. Restoring native communities along highways may have to be a longer-term goal.

PROBLEM: Local conservation organizations and land trusts have to deal with a wide variety of invasive species on their preserves, but sometimes they are unaware of (or unwilling to use) the full range of tools at their disposal. In particular, some organizations might reject biocontrol as an option for reducing the impacts of invasive species because biocontrol itself requires the introduction of nonnative species. Such organizations end up wasting time and money spraying or ripping out invasive plants that could be much more effectively controlled with an available biocontrol agent. For example, a municipal agency that manages a large watershed in New England refused biocontrol researchers permission to release natural enemies of hemlock woolly adelgid, even though this Asian pest is a severe threat to the extensive hemlock stands in the area. In another case, a land conservancy in New

England recently reported in its newsletter that it had mobilized a major volunteer effort to remove purple loosestrife from one of its preserves manually—even though this approach is in the long run completely ineffective, and several ecologically screened biocontrol agents are already available for loosestrife. This approach not only diverts human and financial resources that could be better used on other projects, but it also spreads a distorted view of the relative risks and costs of biological control—and may cause serious ecological damage by blocking the timely use of an effective solution to a pest problem.

WHAT TO DO: Conservation organizations listen to their members. If they start hearing from their members that they should be investigating biological control as a solution to some particular invasive species problem, they are more likely to consider the possibility. Of course, biocontrol is not always the best solution (nor has a program been developed for every pest), but if it is one of the options already available for controlling an invasive species, it ought to be considered.

PROBLEM: Coverage of alien species issues is growing in the media, but the issue has received nowhere near as much coverage as other environmental topics. Though the introduction of nonnative species is widely recognized in the scientific community as one of the four major human impacts on biodiversity and native ecosystems (along with habitat destruction, pollution, and global climate change), it still rates as incidental even for many people who are concerned about the environment.

WHAT TO DO: Exposure is key. If alien species start making the front page and the editorial page and the evening news and the covers of newsletters on a regular basis, the issue will become a standard part of what people think about and are concerned about. Anyone can submit an op-ed or a guest column to a newspaper or suggest a story to a features editor or reporter. The media are always looking for good stories, but what they cover depends largely on what they hear about. Spread the word!

## At State and National Levels

PROBLEM: Some state fish and game agencies still stock rivers with brown trout, rainbow trout, and other fish that are not native to the place where they are released. Other agencies plant invasive shrubs like honeysuckle and autumn olive for bird habitat. Many of these agencies were founded with a single purpose in mind: to improve hunting and fishing opportunities. Broader ecological goals have been added to their missions over the years, but often their fundamental mandates have not changed very much.

WHAT TO DO: Some states have formal mechanisms for incorporating suggestions and ideas from citizens into management practices. For example, the Wisconsin Department of Natural Resources holds large public meetings across the state every year at which suggestions made during the previous twelve months are voted on by everyone who attends and then used as a guide for future management. In any case, fish and game agencies need to hear from people who hunt and fish that introducing nonnative species as game is no longer an acceptable practice and that ecologically sound management for native species is the most appropriate management strategy.

PROBLEM: Many national conservation organizations have begun to pay attention to the threat of alien species introductions, but as a whole, the conservation movement still has a long way to go if it is to address the issue effectively. Given the pace of globalization and therefore of new invasions, a piecemeal approach to this threat is not comprehensive enough to be effective in the long run.

WHAT TO DO: What is needed is a coordinated national campaign to make invasive species threats a major issue. No one organization can do this alone. Conservation organizations have formed coalitions to focus attention on key issues such as pesticide impacts and habitat loss, and the same model could be applied to a campaign to move invasive species onto the national agenda. Such a project is more likely to happen if a wide variety of organizations feel they have a mandate from their members to devote significant resources to the issue.

PROBLEM: Some aspects of the invasive species problem can't be fixed by individuals, but require legislative or regulatory solutions. For example, if the importation of untreated wood is allowed under federal regulations, forest pests are likely to be introduced no matter how many individual people buy only domestic lumber. Unfortunately, there are numerous gaps in the federal laws intended to prevent and manage invasions of nonnative species. These gaps will have to be filled before the system can be called effective, for getting it right even 95 percent of the time isn't good enough when it comes to preventing introductions.

WHAT TO DO: Keep informed about invasive species–related legislation and urge your senators and representatives to sponsor it. Even more effective is to persuade some of the major environmental lobbying groups such as the Sierra Club and the National Audubon Society to make alien species legislation a part of their agenda.

# Resources

Listed below are a few books and Web sites that are useful sources of information on alien species issues.

## *Books*

*The Song of The Dodo,* by David Quammen (New York: Scribner, 1996)

A compelling around-the-world adventure story about biogeography, ecological isolation, and invasive species.

*Life Out of Bounds: Bioinvasion in a Borderless World,* by Christopher Bright and Linda Starke (W.W. Norton and Company, 1998)

An examination of the impacts of the loss of ecological isolation on the movement of species around the globe.

*Harmful Non-Indigenous Species in the United States* (Washington, D.C.: Office of Technology Assessment, U.S. Government Printing Office, 1993)

A government report crammed with facts and interesting tables and graphs that provides a broad overview of the subject.

*Strangers in Paradise,* edited by Daniel Simberloff, Don Schmitz, and Tom Brown (Washington, D.C.: Island Press, 1997)

An extremely detailed source book on invasive species in Florida.

*Biological Pollution: The Control and Impact of Invasive Exotic Species,* edited by B. N. McKnight (Indianapolis: Indiana Academy of Science, 1993)

A series of short chapters highlighting some seventeen cases in enough detail to make them vivid and interesting.

*Plant Invaders,* by Q. C. B. Cronk and J. L. Fuller (London: Chapman and Hall, 1995)

The first few chapters cover the nature of plant invasions, and the remainder of the book provides detailed guidance on chemical and mechanical control methods for some sixty important pest plants. An excellent resource for the land steward.

*Biological Invasions,* by M. Williamson (London: Chapman and Hall, 1996)

A technical treatise on the theory of invasion, including mathematical modeling of the process of range expansion after initial invasion.

*Alien Species in North America and Hawaii: Impacts on Natural Ecosystems,* by George W. Cox (Washington, D.C.: Island Press, 1999)

A comprehensive, readable source of general information on invasive species and their impacts.

*Alien Invasion,* by Robert Devine (Washington, D.C.: National Geographic Press, 1998)

A general overview of alien species impacts on industry, agriculture, and other human endeavors (as well as on natural areas).

## Web Sites

Hawai'i Ecosystems at Risk Project
<http://www.hear.org/>

A comprehensive source of information on a variety of issues related to invasive species in Hawai'i.

Invaders Database System
<http://invader.dbs.umt.edu/>

A compendium of publications, invasion histories, and pest descriptions. Also offers an alert service and a database that can be searched by region as well as by topic. Focuses largely on plants.

The Nature Conservancy's Wildland Weeds Management and Research Program
<http://tncweeds.ucdavis.edu/>

Information on various invaders of wildlands, with a focus on plants.

American Lands Alliance's Invasive Species Page
<http://www.americanlands.org/forestweb/invasive.htm>

Information on current national invasive species policy issues.

California Exotic Pest Plant Council
<http://www.caleppc.org/>

Pest lists and symposia announcements for the state of California.

Alien Plant Working Group of the Plant Conservation Alliance
<http://www.nps.gov/plants/alien/>

Nationwide lists of invaders, factsheets, and links to other sites.

Animal and Plant Health Inspection Service (APHIS)
<http://www.aphis.usda.gov>

Information on all major regulatory issues. Special reports on "hot topics."

# Conclusion

The solution to all problems begins with recognition of the fact that a problem exists.

A quick search of major U.S. newspapers found a total of 320 articles published during the 1990s that mentioned the words "invasive species." Of those articles, more than a third were published in 1999 alone. Invasive species are a centuries-old problem that is finally beginning to receive the attention it deserves.

So who is paying attention?

The federal government clearly is beginning to pay attention. President Clinton's 1999 Executive Order on nonindigenous species was a major step forward. If it is implemented well and fully, it could in the long run be the most important thing President Clinton does in his eight years in office.

You are paying attention. You bought this book and read it cover to cover.

And whoever you talk to next will soon be paying attention—if you tell them about what is going on, and if you work together. If there ever was a time to think big—to think beyond you and me and them to *us*—that time is now. If everyone just tries to save his or her own little piece of the picture, much of what is beautiful in the world will disappear. There is no such thing as an individual solution to a global crisis. We have to do it together.

Enjoy the world, spread the word, and share your strength.

# Background Information on Biological Control in Natural Areas

Table 1. Relative degrees of risk posed by different categories of natural enemies

| Type of Natural Enemy | Risk (10 = high, 1 = low) | Comments |
|---|---|---|
| Predacious birds, mammals, fish, or other vertebrates | 10 | Inherently unsafe; should never be introduced beyond historical range of native distribution. |
| Plant-eating fish | 8 | These are generalists and can cause extensive ecological degradation. Establishment is almost always harmful. Even sterile forms (triploids) in confined, artificial water bodies such as stock ponds and irrigation canals may be dangerous, as 100 percent sterility cannot be guaranteed. |
| Predatory snails | 10 | Some species are generalist predators and may cause extinctions of nontarget snails, particularly on islands with many species of snails that evolved without native predators. |
| Social insects (e.g., ants, yellow jacket wasps) | 10 | Dangerous. These organisms can decimate local insect diversity, especially native ants. |

*continues*

Table 1. *Continued*

| Type of Natural Enemy | Risk (10 = high, 1 = low) | Comments |
|---|---|---|
| Tachinid flies | some 10, some 1 | Risk is highly variable; some species, e.g., *Compsilura concinnata,* are broad generalists that have affected many native species; others, such as *Myiopharus* spp. that attack Colorado potato beetles, are highly specific. Host range screening is needed to identify and reject use of generalists. |
| Plant-eating insects | 1–2 (if properly tested for host range) | Host range screening can be used effectively to identify and avoid species with broad host ranges. Use of species with narrow host ranges has had little effect on non-target species. |
| Plant pathogens (rusts and smuts only) | 1 | Only pathogens with high levels of host specificity can be used. Species in two groups, the rusts and smuts, often have extremely narrow host ranges and are excellent candidates for biological control. Other types of plant pathogens that have wide host ranges are not suitable for use. |
| Insect parasitoids | 1–3 | Host ranges are relatively narrow for many species. Nontarget effects can be limited to groups of insects closely related to the target host if host range screening is used to pick agents with narrow host ranges. |
| Insects that are predators | 3 | Prey ranges are usually wider than parasitoid host ranges, but effects on individual prey species are diluted by lack of specificity. Closely related predators may be affected by competition for resources. |
| Mites that are predators | 1–2 | Prey ranges are variable and most species are not specialists; however, because their prey are extremely numerous, no negative effects have been documented on nontarget species. |

Table 2. Examples of nonnative plants damaging to natural areas against which biological control has been attempted, with emphasis on the United States, South Africa, and Australia

| Area | Plant Species | Affected Community or Ecosystem |
|---|---|---|
| Southwestern United States | saltcedar | Riparian communities along desert streams |
| Florida Everglades | Australian paperbark tree | Sawgrass marshes |
| Northern United States | purple loosestrife | Freshwater marshes, ponds |
| Hawai'i | mist flower | Forests |
| Hawai'i | banana poka | Forests |
| Hawai'i | *Miconia calvescens* | Forests |
| Florida and Louisiana | water hyacinth | Slow-moving rivers, ponds |
| Florida | water lettuce, hydrilla, alligator weed | Slow-moving rivers and ponds in warm climates |
| Oregon, California, New Zealand, Australia | Klamath weed | Native western grasslands |
| Oregon, New Zealand | tansy ragwort | Coastal grasslands |
| South Africa | *Acacia saligna, Hakea sericea, Acacia longifolia* | Fynbos (a species-rich concentration of endemic plants) |
| Australia | prickly pear cacti | Eucalyptus forests, grasslands |
| Australia | giant sensitive plant | Diverse wetlands, forests |
| Australia | floating fern | Tropical lakes and rivers |
| Australia (SW area) | bridal creeper | Various natural habitats |
| Australia | boneseed | Mountainous habitats of southeastern Australia |
| Australia | bitou bush | Coastal areas of subtropical eastern Australia |
| Australia (Queensland) | rubbervine | Range, riparian areas |
| New Zealand, Hawai'i | gorse | Grasslands, beech forests |

Table 3. Partial list of nonnative insects damaging to natural areas against which biological control has been attempted, with emphasis on the eastern and central United States[a]

| Country and Insect | Threatened Native Plant and Biocontrol Outcome |
|---|---|
| United States and Canada: Balsam woolly adelgid | *Threatened*: Balsam fir. *Outcome*: Numerous natural enemy releases failed to suppress the pest. |
| United States: Hemlock woolly adelgid | *Threatened*: Eastern hemlock and Carolina hemlock. Eastern hemlock is being killed over much of its range, and the pest is still spreading. *Outcome*: Project is in progress; predacious ladybird beetles that feed on the pest have been collected in China and Japan, and releases have started in the United States. |
| United States and Canada: Beech scale | *Threatened*: American beech has been killed over most of its range, but persists through resprouting, followed by reinfection. Restoration potential is high if scale could be suppressed. *Outcome*: No project has been organized, but exploratory work is under way to locate the pest's home range in Asia. |
| United States: Gypsy moth | *Threatened*: Oaks (*Quercus* spp.) and various other eastern hardwoods are periodically defoliated, resulting not in species loss, but in extensive tree death and reduction of acorn mast. *Outcome*: A near 100-year series of introductions appears to have culminated in the establishment of a complex of parasitoids and, most importantly, in a fungal pathogen (*Entomophaga maimaiga*) that now reliably suppresses the pest in New England. |
| United States and Canada: Larch casebearer | *Threatened*: Eastern larch and western larch. These larches were defoliated over large areas before this pest was suppressed. *Outcome*: Two parasitoids, *Agathis pumila* and *Chrysocharis laricinellae*, have lowered the pest's density by 98%.[b] Lower pest densities have ended recurrent tree defoliation, and tree growth rates have returned to normal. |
| United States and Canada: Introduced pine sawfly | *Threatened*: Various species of pines, especially white pine. *Outcome*: A series of introduced parasitoids has suppressed the pest in many areas. |
| Canada: Larch sawfly | *Threatened*: Eastern larch, western larch, and alpine larch. *Outcome*: Several strains of the pest were introduced; parasitoids introduced from Europe have suppressed some but not all strains, and more work is needed. |

| | |
|---|---|
| Canada:<br>European spruce sawfly | *Threatened:* Various species of spruce (*Picea* spp.) in boreal forest were repeatedly defoliated over large areas.<br>*Outcome:* In conjunction with parasitoid introductions, a baculovirus was unintentionally introduced and has completely suppressed the pest. |
| Southern United States:<br>Imported fire ants | *Threatened:* This invasive ant has reduced species diversity of native ants and other invertebrates across a wide area of the southern United States.<br>*Outcome:* A complex of phorid flies in the genus *Pseudacteon* appears to suppress this ant in its native range in South America, and introductions have begun.[c] |
| Florida:<br>Bromeliad weevil,<br>*Metamasius callizona* | *Threatened:* Native bromeliads in Florida are being killed by this newly introduced weevil. The extent of plant losses appears quite significant.<br>*Outcome:* Efforts to find parasitoids in the native range in Mexico and Central America have begun. |
| New Zealand:<br>*Vespula* wasps<br>(yellow jackets) | *Threatened:* Consumption of insect honeydew by these wasps is reducing populations of some native forest birds that rely on the same food source.<br>*Outcome:* The specialized parasitoid *Sphecophaga vesparum vesparum* has been introduced and established; evaluation of control is pending.[d] |
| St. Helena (south Atlantic):<br>Ensign scales | *Threatened:* The endemic gumwood tree<br>*Outcome:* The introduction of the ladybird beetle *Hyperaspis pantherina* has suppressed the nonnative scale that was killing the few remaining gumwood trees. Tree mortality from the pest has now ended.[e] |
| Bermuda<br>(mid-Atlantic):<br>Juniper scales,<br>*Insulaspis pallida* and<br>*Carulaspis minima* | *Threatened:* The endemic Bermuda cedar. Most trees of this species have been killed by these two invasive scales. Only small stands survive on the island.<br>*Outcome:* Several predators and parasitoids were introduced but failed to control the pest, in part because the pests were misidentified and in part because the plant exhibited hypersensitivity to pest feeding, resulting in plant death even at low scale densities. |

[a]Except as noted, taken from R. G. Van Driesche et al., *Biological Control of Arthropod Pests of the Northeastern and North Central Forests in the United States: A Review and Recommendations*. U.S. Forest Service FHTET-96-19 (December 1996).

[b]Larch casebearer: see R. B. Ryan, "Evaluation of Biological Control: Introduced Parasites of Larch Casebearer (Lepidoptera: Coleophoridae) in Oregon," *Environmental Entomology* 19 (1990): 1873–1881.

[c]Phorid flies that attack fire ants: See M. R. Orr et al., "Foraging Ecology and Patterns of Diversification in Dipteran Parasitoids of Fire Ants in South Brazil," *Ecological Entomology* 22 (1997): 305–314.

[d]Use of a parasitoid of a nonnative vespid wasp in New Zealand: See H. Moller et al., "Establishment of the Wasp Parasitoid, *Sphecophaga vesparum* (Hymenoptera: Ichneumonidae), in New Zealand," *New Zealand Journal of Zoology* 18 (1991): 199–208.

[e]Use of a ladybird beetle to save the St. Helena gumwood tree: See R. G. Booth et al., "The Biology and Taxonomy of *Hyperaspis pantherina* (Coleoptera: Coccinellidae) and the Classical Biological Control of its Prey, *Orthezia insignis* (Homoptera: Ortheziidae)," *Bulletin of Entomological Research* 85 (1995): 307–314.

# Species Mentioned in the Text
### *(Common names and corresponding scientific names)*

African feather grass (*Pennisetum macrourum*)

African tuliptree (*Spathodea campanulata*)

alligator weed (*Alternanthera philoxeroides*)

alpine larch (*Larix lyallii*)

American beech (*Fagus grandifolia*)

American bullfrog (*Rana catesbeiana*)

American chestnut (*Castanea dentata*)

American elm (*Ulmus americana*)

American holly (*Ilex opaca*)

apple (*Malus pumilla*)

apple maggot fly (*Rhagoletis pomonella*)

ardisia (*Ardisia elliptica*)

Argentine ant (*Linepithema humile*)

ash whitefly (*Siphoninus phyllyreae*)

Asian chestnut (*Castanea mollissima* or *C. crenata*)

Asian longhorned beetle (*Anoplophora glabipennis*)

aspen (*Populus* spp.)

Australian paperbark tree (*Melaleuca quinquenervia*)

Australian pine (*Casuarina equisetifolia*)

autumn olive (*Elaeagnus umbellata*)

balsam fir (*Abies balsamea*)

balsam woolly adelgid (*Adelges piceae*)

banana (*Musa* spp.)

banana poka (*Passiflora mollissima*)

bean shell (*Villosa fablis*)

beech bark disease (caused by *Nectria coccinea* var. *faginata* and *Nectria galligena*)

beech scale (*Crytococcus fagisuga*)

Bermuda cedar (*Juniperus bermudiana*)

big-headed ant (*Pheidole megacephala*)

big-leafed maple (*Acer macrophyllum*)

big-toothed aspen (*Populus grandidentata*)

Bishop pine (*Pinus muricata*)

biting sand flies (flies in the genus *Phlebotomus*, Psychodidae)

bitou bush (*Chrysanthemoides monilifera rotunda*)

blackberry [species invasive in Hawai'i] (*Rubus argutus*)

black birch (*Betula nigra*)

black sand shell (*Ligumia recta*)

blue jay (*Cyanocitta cristata*)

blue tilapia (*Oreochromis aureues*)

eastern bluebird (*Sialia sialis*)

blue-point (*Amblema plicata plicata*)

boneseed (*Chrysanthemoides monilifera monilifera*)

box elder (*Acer negundo*)

Brazilian peppertree (*Schinus terebinthifolia*)

breadfruit (*Artocarpus altilis*)

bridal creeper (*Myrsiphyllum asparagoides*)

brook trout (*Savelinus fontinalis*)

broomsedge (*Andropogon virgnincus*)

brown tree snake (*Boiga irregularis*)

brown trout (*Salmo trutta*)

buckhorn (*Tritogonia verrucosa*)

bur oak (*Quercus macrocarpa*)

bush beardgrass (*Schizachyrium condensatum*)

butternut (*Juglans cinerea*)

cane tibouchina (*Tibouchina urvilleana*)

Carolina hemlock (*Tsuga caroliniana*)

cattail (*Typha* spp.)

cheat grass (*Bromus tectorum*)

cherry (*Prunus* sp.)

collared peccary (*Tayassu tajacu*)

Colorado potato beetle (*Leptinotarsa decemlineata*)

common opossum (Virginia) (*Didelphis marsupialis*)

cottonwood (*Populus* spp.)

cougar (*Felis concolor*)

crazy ant (*Paratrechina fulva*)

creek heel-splitter (*Lasmigona compressa*)

crested wheatgrass (*Agropyron cristatum*)

cylindrical paper shell (*Anodontoides ferussacianus*)

dandelion (*Taraxacum officinale*)

deer-toe (*Truncilla truncata*)

dodder (*Cuscuta* spp.)

dodo (*Raphus cucullatus*)

Douglas fir (*Pseudotsuga menziesii*)

Dutch elm disease (caused by *Ophiostoma* [= Ceratocystis] *ulmi*)

eastern bluebird (*Sialia sialis*)

eastern hemlock (*Tsuga canadensis*)

eastern larch (*Larix laricina*)

eelgrass (*Zostera marina*)

elk toe (*Alasmidonta marginata*)

eucalyptus trees (*Eucalyptus* spp.)

Eurasian watermilfoil (*Myriophyllum spicatum*)

European green crab (*Carcinus maenas*)

European rabbit (*Oryctolagus cuniculus*)

European spruce sawfly (*Gilpinia hercyniae*)

European starling (*Sturnus vulgaris*)

fat mucket (*Lampsilis radiata luteola*)

fawn's foot (*Truncilla donaciformis*)

fennel (*Foeniculum vulgare*)

fire ants (*Solenopsis invicta*)

floating fern (*Salvinia molesta*)

fountain grass (*Pennisetum setaceum*)

fragile heel-splitter (*Potamilus ohiensis*)

fragile paper shell (*Leptodea fragilis*)

Fraser magnolia (*Magnolia fraseri*)

garlic mustard (*Alliaria petiolata*)

giant African snail (*Achatina fulica*)

giant floater (*Anodonta grandis grandis*)

giant hogweed (*Heracleum mantegazzianum*)

giant sensitive plant (*Mimosa pigra*)

ginkgo (*Ginkgo biloba*)

glossy buckthorn (*Rhamnus frangula*)

gorse (*Ulex europaeus*)

gray dogwood (*Cornus racemosa*)

green ash (*Fraxinus pennsylvanica*)

gumwood [tree in St. Helena] (*Commidendrum robustum*)

gypsy moth (*Lymantria dispar*)

hackle-back (a pearly mussel) (*Lasmigona costata*)

Hawai'i creeper (*Paroreomyza montana*)

hawthorn (*Corylus* spp.)

hemlock woolly adelgid (*Adelges tsugae*)

hickory [the tree] (*Carya* spp.)

hickory nut (a pearly mussel) (*Obovaria subrotunda*)

Hilo grass (*Paspalum conjugatum*)

honeysuckle (*Lonicera* x *bella*)

horn shell (*Uniomerus tetralasmus*)

hydrilla (*Hydrilla verticillata*)

ice plant (*Mesembryanthemum crystallinum*)

Indian kahili ginger (*Hedychium gardnerianum*)

jaguar (*Felis onca*)

Japanese barberry (*Berberis thunbergii*)

Japanese knotweed (*Reynoutria japonica*, formerly *Polygonatum cuspidatum*)

Japanese maple (*Acer palmatum*)

kidney shell (*Ptychobranchus fasciolaris*)

Kikuyu grass (*Pennisetum clandestinum*)

Klamath weed (*Hypericum perforatum*)

koa (*Acacia koa*)

kudzu (*Pueraria lobata*)

lady finger (*Elliptio dilatata*)

lake trout (*Salvelinus namaycush*)

larch case bearer (*Coleophora laricella*)

larch sawfly (*Pristiphora erichsonii*)

leafy spurge (*Euphorbia* x *pseudovirgata*)

Lehmann lovegrass (*Eragrostis lehmanniana*)

liliput shell (*Toxolasma parvus*)

llama (*Lama glama*)

London plane tree (*Platanus acerifolia*)

Lombardy poplar (*Populus nigra* 'Italia')

mamane (*Sophora chrysophylla*)

maple-leaf (*Quadrula cylindrica*)

Mediterranean fruit fly (*Ceratitis capitata*)

mesquite (*Prosopis glandulosa*)

miconia (*Miconia calvescens*)

mist flower (*Ageratina riparia*)

moa (*Dinornis maximus* and other species)

mountain laurel (*Kalmia latifolia*)

multiflora rose (*Rosa multiflora*)

muskrats (*Ondatra zibethicus*)

mynah bird (*Acridotheres tristis*)

Nile perch (*Lates* cf. *niloticus*)

northern club shell (*Pleurobema clava*)

northern riffle shell (*Epioblasma rangiana*)

'ohi'a (*Metrosideros* spp.)

Ohio pig-toe (*Pleurobema sintoxia*)

oriental bittersweet (*Celastrus orbiculatus*)

paper pond shell (*Anodonta imbecillis*)

petrels (members of Hydrobatidae)

pimple-back (*Quadrula pustulosa pustulosa*)

pine sawfly (*Diprion similis*)

pink heel-splitter (*Potamilus alatus*)

pocketbook (*Lampsilis ventricosa*)

pohole fern (*Dipazium sandwichianum*)

prickly pear cacti (*Opuntia inermis*, *O. stricta*, and other species)

purple loosestrife (*Lythrum salicaria*)

purple pimple-back (*Cyclonaias tuberculata*)

quaking aspen (*Populus tremuloides*)

Queen Anne's lace (*Daucus carota*)

rabbit's foot (*Quadrula cylindrica*)

rainbow shell (*Villosa iris iris*)

rainbow trout (*Oncorhynchus mykiss*)

red oak (*Quercus rubra*)

robin (*Turdus migratorius*)

rubbervine (*Cryptostegia grandiflora*)

rusty crawdad (*Orconectes rusticus*)

saltcedar (*Tamarix* spp.)

San José scale (*Quadraspidiotus perniciosus*)

sea lamprey (*Petromyzon marinus*)

semaphore cactus (*Opuntia spinosissima*)

silverswords (*Argyroxiphium* spp.)

Simpson's mussel (*Simpsonaias ambigua*)

slipper shell (*Alasmidonta viridis*)

slippery elm (*Ulmus rubra*)

small mucket (*Lampsilis fasciola*)

smooth brome (*Bromus inermis*)

snuffbox (*Epioblasma triquetra*)

squaw foot (*Strophitus undulatus undulatus*)

strawberry guava (*Psidium cattleianum*)

swamp willow (*Decodon verticillatus*)

tanagers (*Piranga* spp.)

tansy ragwort (*Senecio jacobaea*)

tree of heaven (*Ailanthus altissima*)

turkey (*Meleagris gallopavo*)

twice-stabbed ladybird beetle (*Chilocorus stigma*)

Wabash big-toe (*Fusconaia flava*)

washboard (*Megalonaias nervosa*)

water hyacinth (*Eichhornia crasipes*)

water lettuce (*Pistia stratiotes*)

western larch (*Larix occidentalis*)

white ash (*Fraxinus americana*)

white birch (*Betula papyrifera*)

white heel-splitter (*Lasmigona compressa*)

white pine (*Pinus strobus*)

winged loosestrife (*Lythrum alatum*)

wisteria (*Wisteria* spp.)

wood thrush (*Hylocichla mustelina*)

yellow birch (*Betula alleghaniensis*)

yellow-legged frog (*Rana boylii*)

zebra mussel (*Dreissena polymorpha*)

# Endnotes

## Chapter 1

pp. 7–8      Ecological impacts of Polynesian settlement: See L. W. Cuddihy and C. P. Stone, *Alteration of Native Hawaiian Vegetation: Effects of Humans, Their Activities and Introductions* (Honolulu: University of Hawai'i Cooperative National Park Resources Studies Unit, 1990), 17–36.

p. 9      Ecological impacts of European settlement: See Cuddihy and Stone, *Alteration of Native Hawaiian Vegetation*, 37–99.

p. 9      Ecology of introduced animals: See C. Lever, *Naturalized Animals: The Ecology of Successfully Introduced Species* (London: T. & A. D. Poyser, 1994).

p. 9      Vulnerability of the Hawaiian ecosystem to invasion: See D. Simberloff, "Conservation Biology and the Unique Fragility of Island Ecosystems," in W. L. Halvorson and G. J. Maender, eds., *The Fourth California Islands Symposium: Update on the Status of Resources* (Santa Barbara, Calif.: Santa Barbara Museum of Natural History, 1994), 1–10; and B. E. Coblentz, "Exotic Organisms: A Dilemma for Conservation Biology," *Conservation Biology* 4(3): 261–264.

p. 9      Increasing domination of Hawai'i by alien species: See The Nature Conservancy of Hawai'i, *Hawaiian High Islands Ecoregional Plan* (Honolulu: The Nature Conservancy of Hawai'i, 1999).

pp. 9–10      General impacts of feral pigs on native Hawaiian ecosystems: See C. P. Stone, "Alien Animals in Hawaii's Native Ecosystems: Toward Controlling the Adverse Effects of Introduced Vertebrates," in C. P. Stone and J. M. Scott, eds., *Hawaii's Terrestrial Ecosystems: Preservation and Management* (Honolulu: Cooperative National Park Resources Studies Unit, University of Hawaii, 1985), 251–297.

pp. 10–15      History of feral pig invasion into Kipahulu Valley: See C. H. Diong, *Population Biology and Management of the Feral Pig (Sus scrofa L.) in Kipahulu Valley, Maui* (Ph.D. Dissertation, Department of Zoology, University of Hawai'i, 1982).

pp. 10–11    Report of 1945 expedition to Kipahulu: See G. O. Fagerlund and F. A. Hjort, *Kipahulu Valley: Report on Above by Chief Ranger Fagerlund; Description by Ranger Hjort* (United States Department of the Interior, National Park Service, Hawaii National Park, 1945; File No. 0-32).

pp. 11–12    Report of 1967 scientific expedition: See R. E. Warner, ed., *Scientific Report of the Kipahulu Valley Expedition* (Maui, Hawai'i: The Nature Conservancy, 1967).

pp. 12–13    Report of the 1976 expedition: See C. Lamoureux and L. Stemmermann, *Report of the Ki-Pahulu Bicentennial Expedition* (Honolulu: University of Hawai'i Cooperative National Park Resources Studies Unit, 1976).

p. 13    Study of Kipahulu pig population biology: See Diong, *Population Biology and Management*.

p. 13    Park Service interdisciplinary study and pig removal: See S. J. Anderson and C. P. Stone, "Snaring to Control Feral Pigs *Sus scrofa* in a Remote Hawaiian Rain Forest," *Biological Conservation* 63 (1993): 195–201.

p. 14    Difference in recovery between lower and upper units: From an interview with S. J. Anderson, Resources Management Division, Haleakala National Park, Makawao, Hawai'i, 19 July 1999.

p. 16    Reforestation of former pasture with koa: See P. G. Scowcroft and J. Jeffrey, "Potential Significance of Frost, Topographic Relief, and *Acacia koa* Stands to Restoration of Mesic Hawaiian Forests on Abandoned Rangeland," *Forest Ecology and Management* 114 (1999): 447–458.

pp. 16–17    Endemic birds of Hakalau: For a definitive treatment of the native forest birds of Hakalau and of Hawai'i in general, see J. M. Scott et al., *Forest Bird Communities of the Hawaiian Islands: Their Dynamics, Ecology, and Conservation; Studies in Avian Biology No. 9* (Los Angeles: Cooper Ornithological Society, 1986); for a brief overview of conservation issues as they relate to native Hawaiian forest birds, see J. M. Scott et al., "Conservation of Hawaii's Vanishing Avifauna," *BioScience* 38 (4) (1988): 238–252.

p. 17    Spread of mosquitoes and disease in Hakalau: From C. Atkinson and D. LaPointe, personal communication (based on unpublished data and articles in preparation).

p. 19    Impacts of pigs on native Hawaiian forest ecosystems: See Cuddihy and Stone, *Alteration of Native Hawaiian Vegetation*, 64–65; and J. K. Baker, "The Feral Pig in Hawai'i Volcanoes National Park,"in *Proceedings of the First Conference on Scientific Research in National Parks* (New Orleans, 1976), 365–367.

p. 20    Lobeliads at Waikamoi: From an interview with R. Hobdy, Hawai'i Department of Land and Natural Resources, Division of Forestry and Wildlife, Maui, Hawai'i, 13 January 1999.

p. 20    New research emphasis on whole-ecosystem conservation: From an interview with J. Jacobi, Biological Resources Division, U.S. Geological Survey, Volcano, Hawai'i, 2 August 1999.

p. 22      History of forestry and forest reserves: See Cuddihy and Stone, *Alteration of Native Hawaiian Vegetation*, 48–59.

p. 23      'Ohi'a dieback as a natural, cyclical phenomenon: See L. L. Loope and D. Mueller-Dombois, "Characteristics of Invaded Islands, with Special Reference to Hawai'i," in J. A. Drake et al., eds., *Biological Invasions: A Global Perspective*, SCOPE 37, ICSU (Chichester, U.K.: John Wiley and Sons, 1989), 257–280.

p. 23      Relative impacts of pigs as compared with those of humans and other species: See Stone, "Alien Animals in Hawaii's Native Ecosystems," 251–297.

p. 25      Invasion success of birds in the Hawaiian Islands: See M. Williamson and A. Fitter, "The Varying Success of Invaders," *Ecology* 77(6) (1996): 1661–1666.

p. 25      Ecological reasons for severity of nonnative species impacts on islands: See D. Simberloff, "Why Do Introduced Species Appear to Devastate Islands More Than Mainland Areas?" *Pacific Science* 49(1) (1995): 87–97.

p. 25      "Isolated oceanic islands were predisposed . . .": From Loope and Mueller-Dombois, "Characteristics of Invaded Islands, with Special Reference to Hawai'i," 257–280; see also references on the same topic at the beginning of the chapter.

p. 26      Role of Polynesians in making Hawaiian ecosystem vulnerable to invasion: See P. V. Kirch, "The Environmental History of Oceanic Islands," in P. V. Kirch and T. L. Hunt, eds., *Historical Ecology in the Pacific Islands: Prehistoric Environmental and Landscape Change* (New Haven, Conn.: Yale University Press, 1997), 1–21; Cuddihy and Stone, *Alteration of Native Hawaiian Vegetation*; and Loope and Mueller-Dombois, "Characteristics of Invaded Islands, with Special Reference to Hawai'i," 257–280.

p. 26      Nature and numbers of introductions: For a general overview of history of introductions, see D. S. Wilcove, *The Condor's Shadow: The Loss and Recovery of Wildlife in America* (New York: W. H. Freeman and Company, 1999), 211–220; for specific numbers, see Cuddihy and Stone, *Alteration of Native Hawaiian Vegetation*, 63–74.

p. 27      Invasions as triggers for change in ecosystem processes: See C. M. D'Antonio and P. M. Vitousek, "Biological Invasions by Exotic Grasses, the Grass/Fire Cycle, and Global Change," *Annual Review of Ecological Systematics* 23 (1992): 63–87; and Y. Baskin, "Winners and Losers in a Changing World: Global Changes May Promote Invasions and Alter the Fate of Invasive Species," *BioScience* 48(10) (1998): 788–792.

pp. 27–28      Key to success of pig eradication in Hawai'i Volcanoes National Park: From an interview with C. Smith, Department of Botany, University of Hawai'i at Manoa, 17 July 1997.

p. 28      History of pig eradication in Hawai'i Volcanoes National Park: See L. K. Katahira et al., "Eradicating Feral Pigs in Montane Mesic Habitat at Hawai'i Volcanoes National Park," *Wildlife Society Bulletin* 21 (1993): 269–274.

p. 29        Pig hunters in popular press: See D. Nakaso, "Oahu Pig Hunters an Endangered Species," *Honolulu Advertiser*, 8 September 1998, A1; and R. Thompson, "As Preservation Efforts Rise, Hunters Fear Own Extinction," *Honolulu Star-Bulletin*, 27 April 1998, A1.

pp. 29–30   Animal rights campaign against snaring: See The Nature Conservancy of Hawai'i, *The Challenge to Save Hawaii's Forests* (Honolulu: The Nature Conservancy of Hawai'i, 1994); and from various Nature Conservancy of Hawai'i memoranda from Alan Holt to staff and colleagues on the progress of the snaring debate, 1993–94.

p. 30        I went to the PeTA-TNC meeting . . .": As quoted in The Nature Conservancy of Hawai'i, *The Challenge to Save Hawaii's Forests*.

## Chapter 2

p. 33        Extinctions caused by Polynesian settlement of Hawai'i: See P. V. Kirch, "The Impact of the Prehistoric Polynesians in the Hawaiian Ecosystem," *Pacific Science* 36 (1982): 1–14; see also H. F. James et al., "Radiocarbon Dates on Bones of Extinct Birds from Hawaii," *Proceedings of the National Academy of Science* 84 (1987): 2350–2354.

pp. 33–34   Plant and animal introductions to Hawai'i and their effects: See Chapters 8 and 9 in C. P. Stone and D. B. Stone, eds., *Conservation Biology in Hawai'i* (Honolulu: University of Hawai'i Press, 1989).

p. 33        Number of nonnative plant species in Hawai'i: From L. Loope, USGS-BRD, Haleakala Field Station, Maui, Hawai'i, personal communication.

pp. 35–36   Ranges of South American hummingbirds: See J. S. Dunning and R. S. Ridgely, *South American Land Birds* (Newton Square, Pa.: Harrowood Books, 1982).

p. 36        Numbers of species versus latitude for a variety of groups: See M. A. Huston, *Biological Diversity: The Coexistence of Species on Changing Landscapes* (Cambridge, U.K.: Cambridge University Press, 1994); see also M. L. Rosenzweig, *Species Diversity in Space and Time* (Cambridge, U.K.: Cambridge University Press, 1995).

p. 36        Numbers of cichlid species in the Great Lakes of Africa: See G. Fryer and T. D. Iles, *The Cichlid Fishes of the Great Lakes of Africa: Their Biology and Evolution* (Edinburgh: Oliver and Boyd, 1972); for information on species richness of ancient lakes in general (including Lake Baikal), see K. Martens, "Speciation in Ancient Lakes," *Trends in Ecology and Evolution* 12 (1997): 177–182.

p. 36        Ecology, distribution, and conjectured evolution of Darwin's finches: See P. R. Grant, *Ecology and Evolution of Darwin's Finches* (Princeton, N.J.: Princeton University Press, 1986).

p. 37        Map and pictorial presentation of valley-by-valley snail radiations: See S. Carlquist, *Hawaii: A Natural History—Geology, Climate, Native Flora and Fauna Above the Shoreline* (Garden City, N.Y.: The Natural History Press, 1970).

pp. 38–39   Role of Polynesians in extinctions of island birds: See D. W. Steadman, "Extinction of Birds in Eastern Polynesia: A Review of the Record, and

Comparisons with Other Pacific Island Groups," *Journal of Archaeological Science* 16 (1989): 177–205.

p. 40    Limiting the range of attackers: See, for example, C. C. Daehler and D. R. Strong, "Reduced Herbivore Resistance in Smooth Cordgrass (*Spartina alterniflora*) after a Century of Herbivore-free Growth," *Oecologia* 110 (1997): 99–108; and L. Bowen and D. van Varen, "Insular Endemic Plants Lack Defenses Against Herbivores," *Conservation Biology* 11 (1997): 1249–1254.

pp. 40–41    Impact of brown tree snake on the fauna of Guam: See G. H. Rodda et al., "The Disappearance of Guam's Wildlife," *BioScience* 47 (1997): 565–574.

pp. 43–44    Evidence for competitive speciation (also known as "sympatric" speciation): See G. L. Bush, "Sympatric Speciation in Animals: New Wine in Old Bottles," *Trends in Ecology and Evolution* 9 (1994): 285–288.

pp. 44–45    Consequences of hypothetical total loss of macroscale isolation for mammals: See P. M. Vitousek et al., "Biological Invasions as Global Environmental Change," *American Scientist* 84 (1996): 468–478.

pp. 45–46    Numbers of native Hawaiian species and age of the islands: See W. L. Wagner and V. A. Funk, *Hawaiian Biogeography: Evolution on a Hot Spot Archipelago* (Washington, D.C.: Smithsonian Institution Press, 1995); see also D. Mueller-Dombois et al., *Island Ecosystems: Biological Organization in Selected Hawaiian Communities* (Stroudsburg, Pa.: Hutchinson Ross Publishing Co., 1981).

p. 46    Nonnative insects in North America: See R. I. Sailer, "History of Insect Introductions," in C. L. Wilson and C. L. Graham, eds., *Exotic Plant Pests and North American Agriculture* (New York: Academic Press, 1983), 15–38; a computerized database is also available: See L. Knutson et al., "Computerized Data Base on Immigrant Arthropods," *Annals of the Entomological Society of America* 83 (1990): 1–8.

p. 46    Vertebrate introductions worldwide: See C. Lever, *Naturalized Animals: The Ecology of Successfully Introduced Species* (London: T. and A. D. Poyser, Ltd., 1994).

## Chapter 3

p. 49    Early migrations of *Homo sapiens*: See C. B. Stringer and P. Andrews, "Genetic and Fossil Evidence for the Origin of Modern Humans." *Science* 239 (1988): 1263–1268; C. B. Stringer and P. McKie, *African Exodus: The Origins of Modern Humanity* (London: Pimlico, 1997); R. L. Cann, "Hybrids, Mothers, and Clades: Who Is Right?" in T. Akasawa and E. J. E. Szathmary, eds., *Prehistoric Mongoloid Dispersal* (Oxford, U.K.: Oxford Press, 1996), 41–51.

p. 49    Historical pattern of human colonization of the various parts of the world: See D. K. Eliott, *Dynamics of Extinction* (New York: John Wiley and Sons, 1986).

p. 50    Evidence from bones of extinct birds on Pacific Ocean islands: See D. W. Steadman, "Extinctions of Polynesian Birds: Reciprocal Impacts of Birds and People," in P. V. Kirch and T. L. Hunt, eds., *Historical Ecology in the Pacific Islands* (New Haven, Conn.: Yale University Press, 1997), 51–79.

p. 50　　Extinctions following first human invasion of Australia and the Americas: See D. K. Eliott, *Dynamics of Extinction*.

p. 53　　Arrival of the compass in Europe: See A. W. Crosby, *Ecological Imperialism: The Biological Expansion of Europe, 900–1900* (Cambridge, U.K.: Cambridge University Press, 1986).

p. 54　　Evolution of knowledge of wind circulation patterns: See A. W. Crosby, *Ecological Imperialism*, 104–131.

p. 55　　Introduction of *Mayetiola destructor* to North America: See C. L. Metcalf et al., *Destructive and Useful Insects: Their Habits and Control* (New York: McGraw-Hill, 1962), 531.

p. 56　　Release of livestock on islands: See J. L. Chapuis et al., "Alien Mammals: Impact and Management in the French Subantarctic Islands," *Biological Conservation* 67 (1994): 97–104.

p. 56　　Subsequent impacts of livestock left on islands: See A. W. Crosby, *Ecological Imperialism*.

p. 57　　Impacts of cats and rats: See I. A. E. Atkinson, "The Spread of Commensal Species of *Rattus* to Oceanic Islands and Their Effects on Island Avifaunas," in P. J. Moors, ed., *Conservation of Island Birds* (Norwich, England: Page Bros., 1985), 35–81; see also C. R. Veitch, "Methods of Eradicating Feral Cats from Offshore Islands in New Zealand," in *Conservation of Island Birds*.

p. 59　　Spread of Dutch elm disease in North America: See F. T. Campbell and S. E. Schlarbaum, *Fading Forests: North American Trees and the Threat of Exotic Pests* (New York: Natural Resources Defense Council, 1994), 6.

p. 60　　Introduction of European oysters into the Pacific Northwest: See B. Morton, "The Aquatic Species Problem: A Global Perspective and Review," in F. M. D'Iti, ed., *Zebra Mussels and Aquatic Nuisance Species* (Chelsea, Mich.: Ann Arbor Press, 1997), 1–63.

p. 61　　Geographic origin of San José scale: See R. G. Van Driesche et al., *Biological Control of Arthropod Pests of the Northeastern and North Central Forests in the United States: A Review and Recommendations* (Morgantown, W.Va.: U.S. Forest Service, FHTET-96-19, 1996), 28.

p. 61　　Invasion of *Lantana camara* and other plants: See Q. C. B. Cronk and J. L. Fuller, *Plant Invaders* (London: Chapman and Hall, 1995), 20.

p. 64　　Increase in volume of global trade since 1950: See "Trade Winds," *The Economist*, 8 November 1997, 85–89.

p. 64　　Consequences of increase in volume of air cargo: See C. Bright, "Invasive Species: Pathogens of Globalization," *Foreign Policy*, Fall 1999, 50–64.

pp. 64–65　Biological implications of the advent of containerized transport: See Bright, "Invasive Species: Pathogens of Globalization."

p. 65　　Shrimp farming as a source of invasions: See Bright, "Invasive Species: Pathogens of Globalization."

p. 65　　Impacts of plantation forestry and large-scale agriculture: See Bright, "Invasive Species: Pathogens of Globalization."

## Chapter 4

pp. 73–74    Extinction of native mussels as a consequence of zebra mussel invasion: See A. Ricciardi et al., "Impending Extinctions of North American Freshwater Mussels (Unionida) Following the Zebra Mussel (*Dreissena polymorpha*) Invasion," *Journal of Animal Ecology* 67 (1998): 613–619.

pp. 74–76    History, introduction, and spread of zebra mussels in North America: See C. O'Neill et al., "The Introduction and Spread of the Zebra Mussel in North America," *Proceedings of the Fourth International Zebra Mussel Conference, Madison, Wisconsin, March 1994*, 433–446.

p. 76    Biology and ecology of zebra mussels: See entire issue of *American Zoologist* 36 (1996).

p. 76    Growth of zebra mussel populations and impacts on unionids: See D. W. Schloesser et al., "Zebra Mussel Infestation of Unionid Bivalves (Unionidae) in North America," *American Zoologist* 36 (1996): 300–310.

pp. 76–77    Navigation on the Ohio River at time of settlement: See L. R. Johnson, "Engineering the Ohio," in R. L. Reid, ed., *Always a River: The Ohio River and the American Experience* (Bloomington: Indiana University Press, 1991), 180–209.

pp. 77–81    All quotes and information related to modifications of the Ohio: See Johnson, "Engineering the Ohio."

pp. 79–80    Impacts of modification of the Ohio on native mussels: See J. D. Williams et al., "Freshwater Mussels: A Neglected and Declining Aquatic Resource," online at <http://biology.usgs.gov/s+t/frame/f076.htm>.

p. 80    Effects of human-induced stress on ecosystems: See D. J. Rapport, "How Ecosystems Respond to Stress: Common Properties of Arid and Aquatic Systems," *BioScience* 49(3) (1999): 193–203.

p. 81    Data on water quality in Ohio: From water quality information supplied by Ecological Assessment Unit, Ohio Environmental Protection Agency.

pp. 85–86    Formation and accomplishments of Operation Future: See D. W. Hall, *Operation Future: Farmers Protecting Darby Creek and the Bottom Line* (Marysville, Ohio: Ohio State University Extension and Operation Future, no date given); see also J. W. Hopkins, *Does the Darby Creek Project Make Sense? Some Indications from Watershed Farmers* (Marysville, Ohio: Ohio State University Extension, no date given).

## Chapter 5

p. 91    Empirical data supporting the "Tens Rule": See "The Origins and the Success and Failure of Invasions," in M. Williamson, *Biological Invasions* (London: Chapman and Hall, 1996), 28–54.

p. 92    Movement of *Lantana* seeds by nonnative birds: See C. W. Smith, "Non-native Plants," in C. P. Stone and D. B. Stone, eds., *Conservation Biology in Hawai'i* (Honolulu: University of Hawaii Press, 1989), 60–69.

Ants and Community Simplification in Brazil: A Review of the Impact of Exotic Ants on Native Ant Assemblages," in D. F. Williams, ed., *Exotic Ants: Biology, Impact, and Control of Introduced Species* (Boulder, Colo.: Westview Press, 1994), 151–162; see also N. J. Reimer, "Distribution and Impact of Alien Ants in Vulnerable Hawaiian Ecosystems," in *Exotic Ants*, 11–22.

p. 102    Introduction of crazy ants to Colombia: See I. Z. de Polania and O. M. Wilches, "Impacto Ecológico de la Hormiga Loca, *Paratrechina fulva* (Mayr), en el Municipio de Cimitarra (Santander)," *Revista Colombiana de Entomología* 18(1) (1992): 14–22.

p. 102    Invasion of bullfrogs in California: See S. J. Kupferberg, "Bullfrog (*Rana catesbeiana*) Invasion of a California River: The Role of Larval Competition," *Ecology* 78 (1997): 1736–1751.

pp. 102–103 Spread and impacts of rusty crawdads: See C. A. Taylor et al., "Conservation Status of Crayfishes of the United States and Canada," *Fisheries* 21 (1996): 25–38.

pp. 103–104 Impacts of Nile perch on cichlids in Lake Victoria: See T. Goldschmidt et al., "Cascading Effects of the Introduced Nile Perch on the Detritivorous/Phytoplanktivorous Species in the Sub-Littoral Areas of Lake Victoria," *Conservation Biology* 7 (1993): 686–700; see also T. Goldschmidt, *Darwin's Dreampond* (Cambridge, Mass.: The MIT Press, 1996).

pp. 104–105 Impacts of invasive grasses in Hawai'i Volcanoes National Park: See C. M. D'Antonio and P. M. Vitousek, "Biological Invasions by Exotic Grasses, the Grass/Fire Cycle, and Global Change," *Annual Review of Ecology and Systematics* 23 (1992): 63–87.

p. 105    Goats, fire, and species losses in Hawai'i: See C. W. Smith et al., "Impact of Fire in a Tropical Submontane Seasonal Forest," in *Proceedings of the Second Conference on Science and Research in the National Parks, Vol. 10: Fire Ecology, San Francisco, California, 26–30 November 1979* (U.S. National Park Service, 1980), 313–324.

p. 106    Effects of ice plant on soil chemistry: See N. J. Vivrette and C. H. Muller, "Mechanisms of Invasion and Dominance of Coastal Grassland by *Mesembryanthemum crystallinum*," *Ecological Monographs* 47 (1977): 301–318; see also P. M. Kloot, "The Role of Common Iceplant (*Mesembryanthemum crystallinum*) in the Deterioration of Medic Pastures," *Australian Journal of Ecology* 8 (1983): 301–306.

p. 106    Effects of Australian pine on coastal erosion in Florida: See A. Deaton, "Shoreline Monitoring at Long Key," *Resource Management Notes* 6 (1994): 13–14.

## Chapter 6

p. 107    Title of chapter: Used with permission from F. T. Campbell, coauthor of a previous publication of the same name.

p. 107    Effects of various nonnative pests and diseases on eastern trees: See F. T. Campbell and S. E. Schlarbaum, *Fading Forests: North American Trees and the Threat of Exotic Pests* (New York: Natural Resources Defense Council, 1994).

p. 110       Introduction of scale and spread of beech bark disease: See D. R. Houston, "Major New Tree Disease Epidemics: Beech Bark Disease," *Annual Review of Phytopathology* 32 (1994): 75–76.

pp. 111–112 Silvicultural techniques for beech management: See B. S. Burns and D. R. Houston, "Managing Beech Bark Disease: Evaluating Defects and Reducing Losses," *Northern Journal of Applied Forestry* 4 (1987): 28–33.

pp. 112–113 History of spread and perceptions of seriousness of hemlock woolly adelgid: See D. Souto et al., "Past and Current Status of HWA in Eastern and Carolina Hemlock Stands," *Proceedings of the First Hemlock Woolly Adelgid Review* (Morgantown, W.Va.: USFS Forest Health Technology Enterprise Team, 1996), 9–12.

p. 113       Possible toxic effects of hemlock woolly adelgid spittle: See M. S. McClure, *Biology and Control of Hemlock Woolly Adelgid*, Bulletin 851 (New Haven, Conn.: The Connecticut Agricultural Experiment Station, 1987), 4.

p. 113       Longevity of infected hemlock trees: See D. Souto et al., "Past and Current Status of HWA in Eastern and Carolina Hemlock Stands," 11.

p. 116       Songbirds associated with hemlock: From an interview with R. Evans, Delaware Water Gap National Recreation Area, Pennsylvania, 22 June 1998.

p. 116       Diversity of mosses and liverworts in hemlock stands: See J. J. Battles et al., "Quantitative Inventory of Understory Vegetation of Two Eastern Hemlock (*Tsuga canadensis*) Stands in the Delaware Water Gap National Recreation Area," (Milford, Pa.: Delaware Water Gap National Recreation Area, unpublished report, 1997), i.

pp. 116–117 Biodiversity of hemlock streams: From R. Evans, personal communication, 20 July 1998; data from an unpublished study conducted by the USGS Biological Resources Division.

pp. 117–118 Consequences of loss of hemlock: See R. Evans, "A Framework to Document Ecosystem Effects: Experience with Hemlock and the Hemlock Woolly Adelgid," in R. F. Billings and T. E. Nebeker, eds., *Proceedings: North American Forest Insect Work Conference: Forest Entomology: Vision 20:21*, Texas Forest Service publication 160 (Texas A&M University: Texas Forest Service, 1996), 125.

p. 119       Ecological history of beech gaps: See N. H. Russell, "The Beech Gaps of the Great Smoky Mountains," *Ecology* 34 (1953): 366–374.

p. 119       Importance of high-altitude beech stands: See W. Blozan, *Dendroecology of American Beech Stands Infested with Beech Bark Disease: A Comparative Study of Stand Dynamics and Temporal Growth Features* (Resources Management and Science Division, Great Smoky Mountains National Park, no date given), 1–3.

p. 123       Economic role of chestnut: See C. R. Burnham, "The Restoration of the American Chestnut," *American Scientist* 76 (1988): 478; M. F. Cochran, "Back from the Brink," *National Geographic*, February 1990, 142.

p. 123       Resprouting ability of chestnut, historical increase in abundance: See G. G. Whitney, *From Coastal Wilderness to Fruited Plain: A History of*

Congress, Office of Technology Assessment, *Harmful Non-Indigenous Species in the United States*, report prepared by E. Chornesky et al. (Washington, D.C.: GPO, 1993).

pp. 151–152  Incompatibility of World Trade Organization regulations with the prevention of invasions: See P. Goldman and J. Scott, *Our Forests at Risk: The World Trade Organization's Threat to Forest Protection* (Earthjustice Legal Defense Fund, 1999), available online at <www.earthjustice.org>.

pp. 151–152  Three-point action plan for improving prevention at the international level: See C. Bright, "Invasive Species: Pathogens of Globalization," *Foreign Policy*, Fall 1999, 50–58.

## Chapter 8

p. 155  Summary of Nature Conservancy sheep management: See P. Schuyler, "Control of Feral Sheep (*Ovis aries*) on Santa Cruz Island, California," in F. Hochberg, ed., *Third California Islands Symposium: Recent Advances in Research on the California Islands* (Santa Barbara: Santa Barbara Museum of Natural History, 1993), 443–452.

pp. 156–160  General history of Santa Cruz Island: See J. Gherini, *Santa Cruz Island: A History of Conflict and Diversity* (Spokane, Wash.: The Arthur H. Clark Company, 1997).

pp. 156–157  Presettlement vegetation: See R. W. Brumbaugh, "Recent Geomorphic and Vegetal Dynamics on Santa Cruz Island, California," in D. Power, ed., *A Multidisciplinary Symposium on the California Islands* (Santa Barbara: Santa Barbara Museum of Natural History, 1980), 139–158.

p. 157  Native American influences on the Santa Cruz Island ecosystem: See S. Junak et al., *A Flora of Santa Cruz Island* (Santa Barbara: Santa Barbara Botanic Garden in collaboration with the California Native Plant Society, 1995), 27–28.

pp. 157–158  Analysis of the history and impacts of ranching on the Channel Islands: See P. G. O'Malley, "Animal Husbandry on the Three Southernmost Channel Islands: A Preliminary Overview, 1820–1950," in W. L. Halvorson and G. J. Maender, eds., *The Fourth California Islands Symposium: Update on the Status of Resources* (Santa Barbara: Santa Barbara Museum of Natural History, 1994), 157–164.

pp. 157–158  Detailed history of land use on the island: See Junak et al., *A Flora of Santa Cruz Island*, 27–35.

p. 158  Analysis of the relationship between vegetative cover and landslides, both on Santa Cruz Island and in Mediterranean ecosystems in general: See R. Brumbaugh et al., "Effects of Vegetation Change on Shallow Landsliding: Santa Cruz Island, California," *General Technical Report PSW-158* (Berkeley, Calif.: Pacific Southwest Forest and Range Experiment Station, Forest Service, U.S. Department of Agriculture, 1982), 397–402.

p. 158  Shifts in erosion patterns and geomorphic regime: See Brumbaugh et al., "Recent Geomorphic and Vegetal Dynamics on Santa Cruz Island, California."

p. 158      Early sheep control efforts: See Junak et al., *A Flora of Santa Cruz Island*, 34.

p. 158      Bishop pine losses: See E. Hobbs, "Vegetation Dynamics of a California Island," *General Technical Report PSW-158* (Berkeley, Calif.: Pacific Southwest Forest and Range Experiment Station, Forest Service, U.S. Department of Agriculture, 1982), 603.

p. 159      Overall changes in vegetation: See Junak et al., *A Flora of Santa Cruz Island*, 36–41.

p. 159      Ecological impacts of feral sheep and preference for native plants: See D. Van Vuren and B. E. Coblentz, "Some Ecological Effects of Feral Sheep on Santa Cruz Island, California, USA," *Biological Conservation* 41 (1987): 253–268; and L. Bowen and D. Van Vuren, "Insular Endemic Plants Lack Defenses Against Herbivores," *Conservation Biology* 11 (1997): 1249–1254.

pp. 159–160 Population data: See D. Van Vuren and B. E. Coblentz, "Population Characteristics of Feral Sheep on Santa Cruz Island," *Journal of Wildlife Management* 53(2) (1989): 306–313.

p. 160      Exclosure study: See Brumbaugh, "Recent Geomorphic and Vegetal Dynamics on Santa Cruz Island, California." Emphasis added.

p. 160      Nature Conservancy management objectives on Santa Cruz Island: As quoted in P. Schuyler, "Control of Feral Sheep (*Ovis aries*) on Santa Cruz Island, California," in F. G. Hochberg, ed., *Recent Advances in California Islands Research: Proceedings of the Third California Islands Symposium* (Santa Barbara: Santa Barbara Museum of Natural History, 1989), 443–452.

pp. 161–162 Research for and execution of the eradication campaign: Unless otherwise noted, see Schuyler, "Control of Feral Sheep."

pp. 166–167 Invasion ecology of fennel on Santa Cruz Island: See S. W. Beatty and D. I. Licari, "Invasion of Fennel (*Foeniculum vulgare*) into shrub communities on Santa Cruz Island, California," *Madrono* 39(1) (1992): 54–66.

p. 167      Observations on the relationship between past land use and the distribution of fennel were made by Orrin Sage, who supervised the removal of cattle from the island: As cited in B. A. Dash and S. R. Gliessman, "Nonnative Species Eradication and Native Species Enhancement: Fennel on Santa Cruz Island," in Halvorson and Maender, eds., *The Fourth California Islands Symposium*, 505–512.

p. 169      Study of changes in vegetation following sheep removal: In R. C. Klinger et al., "Vegetation Response to the Removal of Feral Sheep from Santa Cruz Island," in Halvorson and Maender, eds., *The Fourth California Islands Symposium*, 341–350.

p. 169      Quote about the impacts of removing a nonnative species and processes to be studied in designing a removal program: From B. Brenton and R. Klinger, "Modeling the Expansion and Control of Fennel (*Foeniculum vulgare*) on the Channel Islands," in Halvorson and Maender, eds., *The Fourth California Islands Symposium*, 497–504.

pp. 170–171 Impacts of feral pigs in general: See P. Kotanen, "Effects of Fetal [sic]

R. G. Lym, "Leafy Spurge Control: 10 Years of Research Enhancement," *North Dakota Farm Research Bimonthly Bulletin* 47(6) (1990): 3–6.

pp. 208–209    Foreign exploration and testing for natural enemies of spurge: See Pemberton, "Leafy Spurge."

p. 209    Introductions to control prickly pear in Australia: See F. Wilson, *A Review of Biological Control of Insects and Weeds in Australia and Australian New Guinea*, Technical Communication No. 1, Commonwealth Institute of Biological Control, Ottawa, Canada (Farnham Royal, Bucks, England: Commonwealth Agricultural Bureaux, 1960).

p. 209    Insects introduced for spurge control: See N. R. Spencer, "Insects for Leafy Spurge Control," paper presented at Leafy Spurge Symposium, Bozeman, Mont., July 1994.

p. 210    Control of ash whitefly: See C. H. Pickett et al., "Establishment of the Ash Whitefly Parasitoid *Encarsia inaron* (Walker) and Its Economic Benefit to Ornamental Street Trees in California," *Biological Control* 6 (1996): 260–272.

p. 210    Control of tansy ragwort: See E. M. Coombs et al., "Economic and Regional Benefits from the Biological Control of Tansy Ragwort, *Senecio jacobaea*, in Oregon," in V. C. Moran and S. H. Hoffman, eds., *Proceedings of the IX International Symposium on Biological Control of Weeds, Stellenbosch, South Africa, 1996*, 489–494.

pp. 211–212    Integration of various spurge control technologies: See R. G. Lym, "Integrated Chemical and Biological Control of Leafy Spurge," in *Proceedings of the Leafy Spurge Strategic Planning Workshop, Dickinson, ND, 29–30 March 1994*; K. G. Beck and J. R. Sebastian, "Progress on an Integrated Leafy Spurge Management System Combining Sheep Grazing with Fall-Applied Herbicides," in *Proceedings of the 1993 Great Plains Agricultural Council Leafy Spurge Task Force Symposium, Silvercreek, CO, 26–28 July 1993*; and D. D. Biesboer et al., "Controlling Leafy Spurge in Minnesota with Competitive Species and Combined Management Practices," in *Proceedings of the 1993 Great Plains Agricultural Council Leafy Spurge Task Force Symposium*.

p. 212    Interactions between flea beetles and sheep grazing: See R. Hansen, "Effects of *Aphthona* Flea Beetles and Sheep Grazing in Leafy Spurge Stands," in *Proceedings of the 1993 Great Plains Agricultural Council Leafy Spurge Task Force Symposium*.

p. 212    Complementarity of biocontrol and chemical control: See R. G. Lym, "Integrated Chemical and Biological Control of Leafy Spurge," in *Proceedings of the Leafy Spurge Strategic Planning Workshop, Dickinson, ND, 29–30 March 1994*.

p. 213    Goals of TEAM Leafy Spurge project: See P. C. Quimby Jr. and L. Wendel, *The Ecological Areawide Management of Leafy Spurge: A Cooperative Demonstration of Biologically Based IPM Strategies*. Grant proposal submitted to the USDA Agricultural Research Service, Areawide Pest Management Program, 1997.

p. 214    Priorities for weed control in 1914: See A. L. Stone, *How to Rid Our*

Farms of Weeds, University of Wisconsin Agricultural Experiment Station, Circular 48, June 1914.

p. 214    Measures to improve awareness of weed threats in 1915: See A. L. Stone, State Seed Inspection and Weed Control: 1914, University of Wisconsin Agricultural Experiment Station, Bulletin 254, 1915.

pp. 214–215  National strategy for weed management in 1990s: See Federal Interagency Committee for the Management of Noxious and Exotic Weeds, Pulling Together: National Strategy for Invasive Plant Management, no date given.

pp. 214–215  Extent of current weed infestations and rate of expansion in North America: See Federal Interagency Committee for the Management of Noxious and Exotic Weeds, Pulling Together.

p. 216    Growth of LandCare movement: See D. T. Briese and D. A. McLaren, "Community Involvement in the Distribution and Evaluation of Biological Control Agents: LandCare and Similar Groups in Australia," Biocontrol News and Information 18(2) (1997): 39N–49N.

p. 216    LandCare participation in biocontrol projects: See Briese and McLaren, "Community Involvement in the Distribution and Evaluation of Biological Control Agents."

p. 217    Reasons for LandCare's success: See A. Campbell, LandCare: Communities Shaping the Land and the Future (St. Leonard's, New South Wales: Allen & Unwin, 1994).

p. 217    Origins and distinctive features of LandCare movement: See Campbell, LandCare.

## Chapter 11

p. 221    Work of Australian acclimatization committees: See E. Rolls, They All Ran Wild (Sydney: Angus and Robertson, 1969).

p. 221    Species of Opuntia brought to Australia: Opuntia inermis de Candolle and Opuntia stricta (Haworth).

p. 222    Reasons for success of prickly pear biocontrol project in Australia: See F. Wilson, Commonwealth Agricultural Bureaux, A Review of the Biological Control of Insects and Weeds in Australia and Australian New Guinea (Reading, U.K.: Laport Gilbert and Co., Ltd., 1960), 51–56.

p. 223    General introduction to biological control: See R. G. Van Driesche and T. S. Bellows Jr., Biological Control (New York: Chapman and Hall, 1996).

pp. 223–224  Biocontrol of prickly pear in the Caribbean: See M. H. Julien, ed., Biological Control of Weeds: A World Catalogue of Agents and Their Target Weeds (Canberra, Australia: CAB International/ACIAR, 1992), 43, 47, and 48.

p. 224    Impacts of Cactoblastis cactorum moth on semaphore cactus in Florida: See D. M. Johnson and P. D. Stiling, "Distribution and Dispersal of Cactoblastis cactorum (Lepidoptera: Pyralidae), an Exotic Opuntia-feeding Moth, in Florida," Florida Entomologist 81 (1998): 12–22; see also D. M. Johnson and P. D. Stiling, "Host Specificity of Cactoblastis cactorum

(Lepidoptera: Pyralidae), an Exotic *Opuntia*-feeding Moth in Florida," *Environmental Entomology* 25 (1996): 743–748.

pp. 224–225    Methods of estimating the host ranges of herbivorous insects: See K. L. S. Harley and I. W. Forno, *Biological Control of Weeds: A Handbook for Practitioners and Students* (Melbourne, Australia: Inkata Press, 1992).

pp. 224–225    Issues related to host range testing of agents for biocontrol of insect pests: See R. G. Van Driesche and M. Hoddle, "Should Arthropod Parasitoids and Predators Be Subject to Host Range Testing When Used as Biological Control Agents?" *Agriculture and Human Values* 14 (1997): 211–226; and D. P. A. Sands, "Evaluating Host Specificity of Agents for Biological Control of Arthropods: Rationale, Methodology, and Interpretation," *Proceedings of Colloquium on Host Selection and Specificity in Xth International Symposium on Biological Control of Weeds*, Bozeman, Montana, U.S.A., July 4–14, 1999 (U.S. Forest Service, Forest Health Technology Enterprise Team, Morgantown, W. Va.).

p. 225    Parasitoid host range testing protocols in New Zealand: See B. I. P. Barratt et al., "Laboratory Nontarget Host Range of the Introduced Parasitoids *Microctonus aethiopoides* and M. *hyperodae* (Hymenoptera: Braconidae) Compared with Field Parasitism in New Zealand," *Environmental Entomology* 26 (1997): 694–702.

pp. 226–227    Extinction of snails on Moorea following the introduction of a predatory snail: See J. Murray et al., "The Extinction of *Partula* on Moorea," *Pacific Science* 42 (1988): 150–153.

p. 228    Biological control of melaleuca: See M. Wood, "Aussie Weevil Opens Attack on Rampant Melaleuca," *Agriculture Research* (December 1997): 4–7.

p. 228    Biological control of *Salvinia molesta:* See P. M. Room et al., "Successful Biological Control of the Floating Weed *Salvinia*," *Nature* 294(5) (1981): 78–80.

p. 228    Biological control of saltcedar: See C. J. Deloach et al., "Biological Control Programme Against Saltcedar (*Tamarix* spp.) in the United States of America: Progress and Problems," in V. C. Moran and J. H. Hoffmann, eds., *Proceedings of the IX International Symposium on Biological Control of Weeds, Stellenbosch, South Africa, 19–26 January 1996*, 253–260.

p. 229    Effects of *Mimosa pigra* on plant communities in Australia: See R. W. Braithwaite et al., "Alien Vegetation and Native Biota in Tropical Australia: The Impact of *Mimosa pigra*," *Biological Conservation* 48 (1989): 189–210; for a possible biological control solution, see M. K. Seier and H. C. Evans, "Two Fungal Pathogens of *Mimosa pigra* var. *pigra* from Mexico: The Finishing Touch for Biological Control of This Weed in Australia?" in Moran and Hoffmann, eds., *Proceedings of the IX International Symposium on Biological Control of Weeds*, 87–92.

p. 229    Difficulties of controlling Japanese knotweed and other clonal plants: See A. Townsend, "Japanese Knotweed: A Reputation Lost," *Arnoldia* 57(3) (1997): 13–19.

p. 230    Biological control of pine tree seeds as a means of slowing invasions by *Pinus* spp.: See V. C. Moran et al., "Biological Control of Alien, Invasive Pine Trees (*Pinus* Species) in South Africa," in X*th International Symposium on Biological Control of Weeds*, Bozeman, Montana, July 4–14, 1999.

p. 230      Success rate for biological control of weeds: See R. E. McFadyen-Crutwell, "Successes in Biological Control of Weeds," in X*th International Symposium on Biological Control of Weeds*.

p. 230      Positive ecological impacts of biocontrol of Klamath weed: See C. B. Huffaker and C. E. Kennett, "A Ten-Year Study of Vegetational Changes Associated with Biological Control of Klamath Weed," *Journal of Range Management* 12 (1959): 69–82.

pp. 230–231 Early history of biocontrol of insects: See P. DeBach and D. Rosen, *Biological Control by Natural Enemies* (Cambridge, U.K.: Cambridge University Press, 1991), 140–148.

p. 231      Success record of biological control of arthropods: See D. J. Greathead and A. H. Greathead, "Biological Control of Insect Pests by Insect Parasitoids and Predators: The BIOCAT Database," *Biocontrol News and Information* 13(4) (1992): 61N–68N.

p. 231      Impacts of zebra mussels on native pearly mussels: See A. Ricciardi et al., "Impending Extinctions of North American Freshwater Mussels (Unionoida) Following the Zebra Mussel (*Dreissena polymorpha*) Invasion," *Journal of Animal Ecology* 67 (1998): 613–619.

p. 231      Invasion of an earthworm-eating land planarian into Europe: See B. Boag et al., "The Potential Spread of Terrestrial Planarians *Artiposthia triangulata* and *Australoplana sanguinea* var. *alba* to Continental Europe," *Annals of Applied Biology* 127 (1995): 385–390.

p. 232      Research on biocontrol of green crabs: From A. Kuris, University of California at Santa Barbara, personal communication.

pp. 232–233 Biological control of vertebrates: See M. S. Hoddle, "Biological Control of Vertebrate Pests," in T. S. Bellows and T. W. Fisher, eds., *Handbook of Biological Control* (San Diego: Academic Press, 1999), 955–974.

p. 232      Use of biocontrol to suppress cats on Marion Island: See P. J. J. van Rensburg et al., "Effects of Feline Panleucopaenia on the Population Characteristics of Feral Cats on Marion Island," *Journal of Applied Ecology* 24 (1987): 63–73.

p. 233      International biocontrol guidelines: See Anonymous, *Expert Consultation on Guidelines for Introduction of Biological Control Agents*, FAO, Rome, Italy, 17–19, 1991 (1992).

p. 234      Search for native range of *Salvinia molesta*: See P. A. Thomas and P. M. Room, "Taxonomy and Control of *Salvinia molesta*," *Nature* 320 (1986): 581–584.

p. 234      Environmental assessment of the likely host range of arthropod parasitoids proposed for introduction into the United States: See, for example, S. D. Porter, "Field Release of the Decapitating Fly *Pseudacteon curvatus* (Diptera: Phoridae) for Biological Control of Imported Fire Ants (Hymenoptera: Formicideae: *Solenopsis*): An Environmental Assessment, April 1999" (contact USDA-ARS, CMAVE, P.O. Box 14565, Gainesville, Florida, 32604, U.S.A.).

pp. 235–236 Establishment rates for various kinds of biocontrol agents: See M. H.

Julien et al., "Biological Control of Weeds: An Evaluation," *Protection Ecology* 7 (1984): 3–25; see also R. W. Hall et al., "Rate of Success in Classical Biological Control of Arthropods," *Bulletin of the Entomological Society of America* 26(2) (1980): 111–114.

pp. 237–242  Overview of loosestrife biocontrol program: See B. Blossey et al., "A Biological Weed Control Programme Using Insects Against Purple Loosestrife, Lythrum salicaria, in North America," in V. C. Moran and J. J. Hoffmann, eds., *IX International Symposium on Weed Biological Control, Stellenbosch, South Africa, 19–26 January 1996* (Cape Town: University of Cape Town, 1996), 353–355.

p. 238  Effects of purple loosestrife on native plants and animals: See D. Q. Thompson et al., *Spread, Impact, and Control of Purple Loosestrife (Lythrum salicaria) in North American Wetlands* (Fish and Wildlife Research 2, United States Department of Interior, Fish and Wildlife Service, Washington, D.C.), 41 pp.

p. 239  Results of host range testing for biocontrol agents for purple loosestrife: See B. Blossey et al., "Host Specificity and Environmental Impact of Two Leaf Beetles (*Galerucella calmariensis* and G. *pusilla*) for Biological Control of Purple Loosestrife (*Lythrum salicaria*)," *Weed Science* 42 (1994): 134–140.

## Chapter 12

pp. 251–252  Early history of horticulture: See J. Fisher, *The Origins of Garden Plants* (London: Constable and Company, Ltd., 1982).

pp. 252–253  Ornamental plant introductions to North America: See S. A. Spongberg, *A Reunion of Trees: The Discovery of Exotic Plants and Their Introduction into North American and European Landscapes* (Cambridge, Mass.: Harvard University Press, 1990).

p. 253  Cultural symbolism of introduced ornamental plants: See C. Lyon-Jenness, "Bergamot Balm and Verbenas: The Public and Private Meaning of Ornamental Plants in the Mid-Nineteenth-Century Midwest," *Agricultural History* 73(2) (1999): 201–221.

p. 253  Compendium of origins and impacts of invasive plants: See J. M. Randall and J. Marinelli, eds., *Invasive Plants: Weeds of the Global Garden*, Twenty-First Century Gardening Series, no. 149 (Brooklyn, N.Y.: Brooklyn Botanic Garden, 1996).

pp. 254–255  Distribution and abundance of buckthorn and honeysuckle: See L. Li, *The Relationship between Native Woody Plants of Woodlands and Environmental Factors in an Urban Area: A Case Study of Madison, Wisconsin*. Masters Thesis, Department of Landscape Architecture, University of Wisconsin–Madison, 1994.

pp. 255–256  Competitive advantages that make honeysuckle and buckthorn invasive: See R. Harrington et al., "Ecophysiology of Exotic and Native Shrubs in Southern Wisconsin," *Oecologia* 80 (1989): 356–367.

p. 256  Impacts of honeysuckle and buckthorn on songbird nesting success: See K. A. Schmidt and C. J. Whelan, "Effects of Exotic *Lonicera* and *Rham-*

*nus* on Songbird Nest Predation," *Conservation Biology* 13(6) (1999): 1502–1506.

p. 256    Buckthorn and honeysuckle as a largely urban problem: See V. M. Kline and T. McClintock, "The Ground Layer of an Oak Forest in Transition under Prescribed Burning," *Proceedings of the 1993 Midwest Oak Savanna Conference*, online at <http://www.epa.gov/docs/grtlakes/oak/oak93/kline.htm>.

pp. 256–258    Factors contributing to vulnerability of urban ecosystems: See A. M. Davis and T. F. Glick, "Urban Ecosystems and Island Biogeography," *Environmental Conservation* 5 (1978): 299–304.

pp. 259–260    Management plan for the Campus Natural Areas: See V. Kline and B. Bader, *UW–Madison Campus Natural Areas Management Plan* (Madison, Wisc.: UW–Madison Arboretum, 1996).

pp. 260–262    Perspectives on the meaning of "natural": See E. Katz, *Nature as Subject: Human Obligation and Natural Community* (London: Rowman and Littlefield Publishers, 1997); B. McKibben, *The End of Nature* (New York: Anchor Books, 1989); K. Soper, *What Is Nature? Culture, Politics, and the Non-Human* (Oxford and Cambridge: Blackwell Publishers, 1995); M. E. Soulé and G. Lease, *Reinventing Nature? Responses to Postmodern Deconstruction* (Washington, D.C.: Island Press, 1995); and William Cronon, ed., *Uncommon Ground: Rethinking the Human Place in Nature* (New York: W.W. Norton, 1996).

p. 264    "The turning point came . . .": From B. J. Bader and D. Egan, "Community-Based Ecological Restoration: The Wingra Oak Savanna Project," *Orion Afield*, Spring 1999: 30–33.

pp. 265–266    "Those who regularly attend . . .": From Bader and Egan, "Community-Based Ecological Restoration."

p. 266    Perspectives on the ethics and the practicalities of restoration: See R. Elliot, *Faking Nature: The Ethics of Environmental Restoration* (London and New York: Routledge Environmental Philosophies Series, 1997); A. D. Baldwin Jr. et al., eds., *Beyond Preservation: Restoring and Inventing Landscapes* (Minneapolis: University of Minnesota Press, 1994); and W. R. Jordan III et al., eds., *Restoration Ecology: A Synthetic Approach to Ecological Research* (Cambridge: Cambridge University Press, 1987).

p. 268    "Generated edge" model: See C. M. Schonewald-Cox and J. W. Bayless, "The Boundary Model: A Geographical Analysis of Design and Conservation of Nature Reserves," *Biological Conservation* 38 (1986): 305–322.

pp. 268–269    Pilot "social buffers" project in Arboretum: See S. Glass and D. Egan, "Watershed Volunteers: An Approach to Dealing with Cross-Boundary Influences," *Yahara Watershed Journal* (1996): 17–21.

## Chapter 13

p. 273    Survey of public opinion regarding alien invasive species issues: See The Kitchens Group, *Issue Analysis: Hawai'i Statewide*. Prepared for The Nature Conservancy, October 1996.

p. 275    Full text of background study on alien species invasions in Hawai'i: See The Nature Conservancy of Hawai'i and the Natural Resources Defense

Council, *The Alien Pest Species Invasion in Hawaii: Background Study and Recommendations for Interagency Planning*, July 1992.

p. 275    Role of The Nature Conservancy in formation and development of CGAPS: From interviews with Grady Timmons, Communications Director, The Nature Conservancy of Hawai'i, on 16 July 1997 and 6 January 1999.

p. 276    Goals of CGAPS public awareness campaign: From "Public Awareness section of CGAPS strategic plan," 30 July 1997 draft, The Nature Conservancy of Hawai'i.

pp. 276–277    Silent Invasion campaign booklet: See Coordinating Group on Alien Pest Species, *The Silent Invasion*, 1996.

p. 277    Importance of expanding focus of Silent Invasion campaign beyond impacts on biodiversity: From an interview with Alan Holt, director of the Hawai'i chapter of The Nature Conservancy, 24 July 1997.

p. 277    General history of miconia invasion and control in Hawai'i: See P. Conant, A. C. Medeiros, and L. L. Loope, "A Multiagency Containment Program for Miconia (*Miconia calvescens*), an Invasive Tree in Hawaiian Rain Forests," in J. O. Luken and J. W. Thieret, eds., *Assessment and Management of Plant Invasions* (New York: Springer, 1997), 249–251.

p. 280    Impacts of miconia in Tahiti: See J. Y. Meyer and J. Florence, "Tahiti's Native Flora Endangered by the Invasion of *Miconia calvescens* DC (Melastomataceae)," *Journal of Biogeography* 23 (1996): 775–781.

p. 280    "... the one plant that could really destroy the native Hawaiian forests": As quoted in A. C. Medeiros et al., "Status, Ecology, and Management of the Invasive Plant, *Miconia calvescens* DC (Melastomataceae) in the Hawaiian Islands," *Bishop Museum Occasional Papers (Records of the Hawaii Biological Survey for 1996)—Part 1, Articles* 48 (1997): 23–36.

pp. 280–281    Ecology of miconia: See J. Y. Meyer, "Observations on the Reproductive Biology of *Miconia calvescens* DC (Melastomataceae), an Alien Invasive Tree on the Island of Tahiti (South Pacific Ocean)," *Biotropica* 30(4) (1998): 609–624.

pp. 281–282    Early history of miconia invasion in Hawai'i: See A. C. Medeiros et al., "Status, Ecology, and Management of the Invasive Plant, *Miconia calvescens* DC (Melastomataceae) in the Hawaiian Islands," 23–36.

pp. 292–293    Impacts of brown tree snake on Guam: See T. H. Fritts and G. H. Rodda, "Alien Snake Threatens Pacific Islands," *Endangered Species Bulletin* 23(6) (1998): 10–11.

p. 293    Potential impacts of brown tree snake on Maui birds: From a memo written by L. Loope to the superintendent of Haleakala National Park, 17 May 1996.

p. 294    FAA mandate: See United States Department of Transportation, *Record of Decision for the Proposed Master Plan Improvements at Kahului Airport, Kahului, Maui, Hawaii* (Federal Aviation Administration, Western-Pacific Region, Hawthorne, Calif.: 26 August 1998).

# About the Authors

JASON VAN DRIESCHE grew up in the Berkshire Hills of western Massachusetts, where he has been climbing trees for as long as he can remember. He went to school at Bard College in New York's Hudson Valley, and graduated in 1993 with a B.A. in Community, Regional, and Environmental Studies. From 1994 to 1996, he served as an Environmental Education volunteer with the Peace Corps in Panama. He has also worked as a land steward for The Nature Conservancy of Kentucky and as ship's cook on the Hudson River Sloop *Clearwater*. He now lives in Madison, Wisconsin, where he is a graduate student in the Institute for Environmental Studies at the University of Wisconsin–Madison. He is Roy's son.

ROY VAN DRIESCHE grew up in the Willamette Valley in western Oregon. He has been collecting insects since he was about twelve. He earned a B.S. in entomology from Oregon State University in 1970 and a Ph.D. in entomology from Cornell in 1975. He settled in western Massachusetts in 1976, where he teaches and researches biological control at the University of Massachusetts–Amherst. His interests center on the role of natural enemies in suppressing insect pests, both in agricultural settings and in natural areas. He is co-author of the textbook *Biological Control*. He is married and has three children.

# Index